Philosophy of Science

ROYAL INSTITUTE OF PHILOSOPHY SUPPLEMENT: 61

EDITED BY

Anthony O'Hear

CAMBRIDGE
UNIVERSITY PRESS

PUBLISHED BY THE PRESS SYNDICATE OF THE UNIVERSITY OF CAMBRIDGE
The Pitt Building, Trumpington Street, Cambridge, CB2 1RP,
United Kingdom

CAMBRIDGE UNIVERSITY PRESS
The Edinburgh Building, Cambridge CB2 8RU, United Kingdom
32 Avenue of the Americas, New York, NY 10013–2473, USA
477 Williamstown Road, Port Melbourne, VIC 3207, Australia
Ruiz de Alarcón 13, 28014 Madrid, Spain
Dock House, The Waterfront, Cape Town 8001, South Africa

Printed in the United Kingdom at the University Press, Cambridge
Typeset by Michael Heath Ltd, Reigate, Surrey

Library of Congress Cataloguing-in-Publication Data applied for

Contents

Preface

This volume consists of the lectures given in the Royal Institute of Philosophy's annual lecture series in London in 2005-06. Both the topics and the contributors testify to the liveliness of the philosophy of science currently, and to the range of interests within the area. They also testify to the ways in which issues and positions in the philosophy of science bear on central questions of philosophy more generally, such as realism, natural kinds and epistemological progress.

I would like to thank all the contributors to the volume for their lectures and articles, and also Marcela Herdova for her editorial assistance and the preparation of the index.

Anthony O'Hear

Preface

List of Contributors

Hasok Chang, University College London

Robin Le Poidevin, Leeds University

Donald Gillies, University College London

Peter Lipton, University of Cambridge

Sophie R. Allen, University of Oxford

John Worrall, The London School of Economics

James Robert Brown, University of Toronto

James Ladyman, Bristol University

Ian Hacking, University of Toronto and College de France

James Garvey, The Royal Institute of Philosophy

Adolf Grünbaum, University of Pittsburgh

Scientific Progress: Beyond Foundationalism and Coherentism[1]

HASOK CHANG

1. Scientific progress and epistemic justification

Scientific progress remains one of the most significant issues in the philosophy of science today. This is not only because of the intrinsic importance of the topic, but also because of its immense difficulty. In what sense exactly does science makes progress, and how is it that scientists are apparently able to achieve it better than people in other realms of human intellectual endeavour? Neither philosophers nor scientists themselves have been able to answer these questions to general satisfaction.

For conveying the importance of scientific progress as something we need to understand, the person I always look to is not a philosopher, but a historian of science: namely George Sarton, who has been called the 'father' of the academic discipline of history of science. In the preface to his monumental *Introduction to the History of Science*, Sarton announced:

> no history of civilization can be tolerably complete which does not give considerable space to the explanation of scientific progress. If we had any doubts about this, it would suffice to ask ourselves what constitutes the essential difference between our and earlier civilizations. Throughout the course of history, in every period, and in almost every country, we find a small number of saints, of great artists, of men of science. The saints of to-day are not necessarily more saintly than those of a thousand years ago; our artists are not necessarily greater than those of early Greece; they are more likely to be inferior; and of course, our men of science are not necessarily more intelligent than those of old; yet one thing is certain, their knowledge is at once more extensive and more accurate. *The acquisition and systematization of positive knowledge is the only human activity which is truly cumulative and progressive.* Our civilization is

[1] This is an updated and expanded version of the lecture given in the Royal Institute of Philosophy seminar series in London on 28 October 2005.

essentially different from earlier ones, because our knowledge of the world and of ourselves is deeper, more precise, and more certain ...[2]

For Sarton, and many others in the general tradition of the Enlightenment, scientific progress was also inextricably linked with social and political progress. In his autobiography, the behaviourist psychologist B. F. Skinner recalled attending Sarton's lectures on the history of science at Harvard University in the early 1930s:

> Sarton was convinced that the world was also moving cumulatively in the direction of a better moral order, and to make his point he described the public torture of the demented [Robert-François] Damiens, whose misfortune it had been to attack the King of France [Louis XV, in 1757] with a pocket knife. Sarton gave us all the gruesome details: the platform in the public square, the huge crowd, the royal family with their children on a balcony, window space sold at a high premium, and then the stages of the torture, the man's hand plunged into a bowl of burning sulfur, time taken to allow him to recover when he fainted, and at last his limbs attached to four horses and torn from his body. "Gentlemen", Sarton said, "we have made progress".

To this recollection Skinner dryly added a brief parenthetical remark: 'Hitler would become Chancellor of Germany a year and a half later.'[3]

To the casual observer, it may seem that Sarton was correct about the inexorable progress of science, though he may have been too optimistic about the progress of society. The philosopher of science who has thought about the question carefully can agree with this view only with considerable difficulty and unease. One of the reasons for the philosopher's difficulty is that the question of scientific progress is intimately linked with the question of epistemic justification, which is itself a very tricky issue. Progress means that something is getting better. If scientific knowledge is getting better, it must mean that our current scientific beliefs are more justified than our previous beliefs (or at least that we have a larger number of beliefs, which are as justified as the beliefs we used to hold). Therefore, in order to *judge* progress we need to be

[2]   G. Sarton, *Introduction to the History of Science*, I (Baltimore: Carnegie Institution of Washington, 1927), 3–4; emphasis original.
[3]   B. F. Skinner, *The Shaping of a Behaviorist* (New York: Alfred A. Knopf, 1979), 57.

able to assess the degree to which beliefs are justified. In order to *make* progress, we should know how to create better-justified beliefs.

In the epistemology of justification, there are two main theories: foundationalism and coherentism. Both have had their influences in the debates on scientific progress. Historically speaking, foundationalism was the doctrine more readily enlisted in the enterprise of understanding and making scientific progress: if we can build a firm foundation of knowledge, then the rest of science is a straightforward process of building on that foundation. Particularly in the seventeenth century, there was a prevalent dream shared by a broad spectrum of European thinkers ranging from Descartes to Bacon: sweep away unreliable systems of knowledge inherited from the past, propped up by authority and tradition; find elements of knowledge that are free from doubt; and build up a new system on the basis of the secure elements. The rationalists attempted to find the correct first principles, and deduce the rest of knowledge from these principles. This enterprise is generally considered to have failed when it came to the development of empirical science, and I will not go into that history here.

The empiricist variety of the foundationalist project had a longer life in science: science would progress by continually accumulating secure facts gained through observations untainted by groundless assumptions, and building up theories by generalization from the facts. But this vision, too, has faded, most importantly through the recognition of the theory-ladenness of observation: since observations rely on theoretical assumptions, observations are only as reliable as the theories involved in making them. If we were to seek observations that do not embody theoretical interpretations, we are reduced to the level of 'sense-data': not anything like 'The photon emitted by a distant star has the wavelength of 8000 angstroms', not even 'I see a red star', but 'Red here now', whose desperate incorrigibility is entirely useless for building scientific knowledge.

Although the theory-ladenness of observation is usually associated with the post-positivist trend in philosophy of science starting around 1960 and most famously represented by the works of Hanson, Kuhn, Feyerabend and the like, the basic problem was already noted by the logical positivists. The Vienna Circle itself was split by the so-called protocol-sentence debate, in which the leading positivists differed sharply over the privileged epistemic status of observational statements. In the course of that debate, Otto Neurath made famous a metaphor against foundationalism, which we now remember as 'Neurath's boat':

3

We are like sailors who have to rebuild their ship on the open sea, without ever being able to dismantle it in dry-dock and reconstruct it from the best components.[4]

Even Moritz Schlick, Neurath's main opponent in the debate, acknowledged that pure observation statements could not serve as a foundation on which to build knowledge, since 'they are already gone at the moment building begins.'[5] Kuhn and Feyerabend only took all these thoughts to their logical conclusion, when they concluded that there could not be any completely decisive and objective empirical testing of theories.

What we are left with after these assaults on foundationalism is coherentism, which is a doctrine that the only form of epistemic justification we can ultimately have is the coherence of a belief with all the other beliefs we hold. This is the idea that Neurath's boat metaphor expresses so well: our task in making scientific progress is to plug up the holes in the leaky boat; what that amounts to is making the various planks in the boat fit together well with each other. But that image is simply inadequate for capturing the nature of scientific progress, which is not just a matter of making our beliefs more consistent with each other. Nor is that simply an inadequacy of Neurath's metaphor. The problem is with the coherentist doctrine itself, which can only understand justification in terms of consistency. This problem of coherentism manifests itself most acutely as the threat of relativism: if we follow the standard coherentist doctrine, any internally consistent system of knowledge is equally justified. One of the greatest consequences of accepting such relativism is that we can no longer make sense of progress as we know it, because there will be no grounds for saying that we have done better by replacing one system of knowledge with another, as long as both are internally consistent. For some, this renunciation of the idea of progress is only an honest response to the poverty of foundationalism. But I think we can do better than that, because foundationalism and coherentism are not the only alternatives. As I will try to explain in the rest of this paper, I

[4]   O. Neurath, 'Protocol Statements (1932/33)', *Philosophical Papers 1913–1946*, R. S. Cohen and M. Neurath (eds.) (Dordrecht: Reidel, 1983), 91–99, on 92.

[5]   M. Schlick, 'On the Foundation of Knowledge [1934]', *Philosophical Papers*, II (1925–1936), H. L. Mulder and B. F. B. van de Velde-Schlick (eds.) (Dordrecht: Reidel, 1979), 370–387, on 382.

think we can best make sense of scientific progress by transcending the false dichotomy and dilemma between foundationalism and coherentism.

2. The problem, and the solution, in brief

Before I get on to the details, I will give a more concise and definitive statement of the problem, and sketch very briefly the solution I want to offer. The root of the problem is that foundationalism is inadequate as an account of how scientific knowledge is justified. I am using a standard definition of foundationalism here: 'According to foundationalists, epistemic justification has a hierarchical structure. Some beliefs are self-justifying and as such constitute one's evidence base. Others are justified only if they are appropriately supported by these basic beliefs.'[6] The problem is that we have failed to find enough self-justifying propositions that are useful in building scientific knowledge. As most of the apparently self-evident propositions are shown to be really in need of justification, we find ourselves mired in an infinite regress. Metaphorically, we end up like the ancient people who reportedly explained that the ground was firm because it rested on the back of large elephants; the elephants stood on a very large turtle. And what about the turtle?

The only obvious alternative to foundationalism is coherentism. Again, I use a standard definition: 'Coherentists deny that any beliefs are self-justifying and propose instead that beliefs are justified in so far as they belong to a system of beliefs that are mutually supportive.'[7] But coherentism, as normally understood, gives no clear picture of how scientific progress can be made. The solution I offer is called *progressive coherentism*.[8] This doctrine is coherentist in the sense that it rejects the search for an absolutely firm foundation. At the same time, it recognizes the possibility of making progress by first accepting a certain system of knowledge without ultimate justification, and then using that system to launch lines of inquiry which can in the end refine and correct the initially affirmed system.

[6]   R. Foley, 'Justification, Epistemic', *Routledge Encyclopedia of Philosophy*, V, E. Craig (ed.) (London: Routledge, 1998), 158–159.
[7]   *Ibid.*, 157.
[8]   For a more detailed exposition of this doctrine, see H. Chang, *Inventing Temperature: Measurement and Scientific Progress* (New York: Oxford University Press, 2004), chapter 5.

A metaphor will be helpful in generating a vivid image of progressive coherentism. Recall the foundationalist metaphor of progress as building on a firm ground [**Figure 1**]. This metaphor is only as good as the flat-earth cosmology which it presumes. What if we modernize this metaphor? The real situation, according to our current view of the universe, is that we build outward on a round earth, not upward on a flat earth [**Figure 2**]. The round-earth metaphor graphically demonstrates the futility of seeking the absolutely firm ground, the ultimate epistemic justification. The earth is not firmly fixed to anything at all; still, we build on the earth because it is a large, dense body that attracts other objects, and we happen to live on it. The round-earth metaphor also nicely suggests that even in the absence of an absolute foundation, there is a sense of grounding, and a clear directionality according to which building needs to take place.

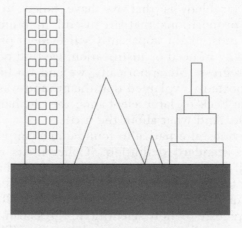

Figure 1. The foundationalist metaphor of building on a firm ground.

3. Progressive coherentism in action (I): justifying and improving measurement methods

3.1. From sensation to instrumentation

Leaving behind the abstractions and metaphors now, I would like to illustrate how progressive coherentism works through some concrete examples. I will present two classes of examples. The first class concern the justification and improvement of measurement

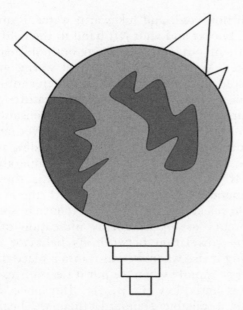

Figure 2. The progressive coherentist metaphor of building on a round earth.

methods. There are several reasons for this focus on measurement. First of all, quantitative measurement is an integral part of modern science, so if we understand how progress is made there, we will understand one important aspect of scientific progress. Secondly, measurement is an area in which science has undoubtedly made unequivocal and lasting progress, so it is an important area to look at when we try to understand progress. (In contrast, there is a sense in which grand theories come and go, and it is very uncertain business debating whether we are approaching the Truth or not.) Thirdly, and perhaps most significantly, I want to discuss the justification of measurement methods because that is where the fatal circularity of empiricist foundationalism is most clearly laid bare for all to see. The examples to be discussed in this section arise from the research I did on thermometry, much of which ended up in my book *Inventing Temperature*.

As an initial example of scientific progress in the absence of indubitable foundations, consider how thermometers can correct and overrule our sensations of hot and cold, much as the initial justification for their adoption had to be based on their agreement with the senses. There is a simple experiment designed to illustrate the unreliability of our temperature sensation. Take three buckets, each filled with hot, cold, and lukewarm water. Plunge your right hand in the hot bucket and your left hand in the cold bucket. After a while, take them out and put them both immediately in the lukewarm bucket. Your left hand will feel warm, and your right hand will feel cold. But inserting a thermometer into various parts of the lukewarm water reveals that the temperature in that body of water is quite uniform. We conclude that our sensations are faulty.

But why do we trust the thermometer, over and above the testimony of our own senses? This is quite puzzling when we think about how we first agreed to measure temperature with the thermometer. If a device we invented with the intention of measuring temperature disagreed wildly with the evidence of our senses, we would reject it. Our initial confidence in something like a column of mercury as a trustworthy indication of temperature comes from the correlation between its behavior and our own sensations: seeing it rise when we put it into a place that feels warm to us, seeing it rise rapidly when we put it near a fire, seeing it drop when we wet it and blow on it, etc. But once we adopt the thermometer as a reliable standard, then we begin to use its readings not only to confirm but also to correct our own sensations.

So here we start with the unjustified initial assumption that our sensations of hot and cold are basically accurate indications of temperature. Then, on the basis of sensations, we then adopt a certain mechanical device to measure temperatures more reliably and precisely than our sensations can. The relation between our sensation and the thermometer is quite subtle, based on what I call the *principle of respect*. Our use of the instrument is made with a respect for sensation as a prior standard, but that does not mean that the verdict of sensation has unconditional authority.

The actual history of the development of simple thermometers shows three main stages.[9] Initially the judgment of temperature was only qualitative, based on the sensation of hot and cold. Then came what we call thermoscopes, instruments that indicate the increase and decrease of temperatures without attaching numbers to them; certified by their broad agreement with sensations, thermoscopes allowed a decisive and consistent comparison and ordering of various phenomena possessing different temperatures. Afterwards numerical thermometers, arising from thermoscopes, went further by attaching meaningful numbers to the degrees of temperature. In this developmental process temperature evolved

[9] For further detail on these three stages of development, see *ibid.*, 47–48.

from an unquantified property to an ordinal quantity, then to a cardinal quantity. This process of development constituted the *creation* of the quantitative concept of temperature, because each stage built on the previous one, but added a new dimension to it.

3.2. Increasing precision

Once a quantitative concept is created, science does not stop there. There is a continual effort to increase precision. To illustrate that point, let me take a more contemporary example—from the 25th of June in 2004, to be precise. I was doing some work then in a teaching lab in the Chemistry Department at University College London, trying to replicate some curious old experiments. One of the things I was investigating was a report by Joseph-Louis Gay-Lussac in 1812 that the temperature of boiling water in a glass vessel was 1.2°C higher than the boiling temperature in a metallic vessel.[10] When I tried this out, I did observe a difference, though it seemed to be only about 0.7–0.8°C according to the common mercury thermometer I was using, graduated down to 1°C.

When I showed this to Andrea Sella, my sponsor at the Chemistry Department, he said 'You need an ancient thermometer', and pulled out a curious old instrument called a Beckmann thermometer, which was in common use not so long ago in physical chemistry for the precise determination of boiling points. The Beckmann thermometer is a thing of beauty, a mercury thermometer with an enormous 6-degree scale graduated down to 0.01°C. One can adapt this instrument to almost any particular temperature range normally desired in a chemistry lab, because the zero point can be moved around by shifting mercury back and forth between the main reservoir at the bottom and an auxiliary reservoir at the top.

I did manage to calibrate the Beckmann thermometer in the range I needed, setting its zero at about 98°C. But then the question presented itself: how do I know what exactly the zero point on the Beckmann thermometer indicates, and how do I know whether the instrument is operating correctly anyway? The only clear guide I had was comparisons with the ordinary mercury thermometer, which I can read at best to about 0.1°C, probably only one-fourth of a degree if I want to be safe. So the ordinary mercury

[10]   Reported in J. B. Biot, *Traité de physique expérimentale et mathématique* (Paris: Deterville, 1816), vol. 1, 42–43.

thermometer can only tell me that the 0.00° on the Beckmann thermometer is, say, about 97.2°C, but not whether it is 97.24°C or 97.27°C; it also cannot tell me whether the behaviour of the Beckmann thermometer is linear between 97.2°C and 97.4°C (which is about a whole centimeter on the Beckmann scale). My trust in the precise behavior of the Beckmann thermometer must come from other sources, although my *basic* trust in it comes from its consistency with the ordinary mercury thermometer. And interestingly, once I gain enough trust in the Beckmann thermometer, I can use its readings not only to refine the readings of the ordinary mercury thermometer, but occasionally to correct or overrule them.

The situation, then, is quite similar to the first story. The relation between the Beckmann thermometer and the ordinary mercury thermometer is like the relation between a thermometer and human sensation. In each case, the prior standard provides initial justification for the later standard, but the later standard refines and even corrects the prior standard.

The case of the Beckmann thermometer shows a bit more clearly the exact mechanism of progress. If we think again about why we trust the Beckmann thermometer beyond where the ordinary mercury thermometer can give us the verdict, we can see how there are other criteria that guide our judgements. First of all, it seems that higher precision in itself is an attraction. This is like how we are predisposed in vision to take the sharper image as the superior picture; when we put on our glasses and see that objects are seen with a sharper focus, we believe that we are getting a better picture; we make that judgement without understanding how the glasses produce a truer picture according to the laws of optics and the physiology of the eye. But of course the higher precision has to be backed up by self-consistency; I would not trust the Beckmann thermometer if it gave me varying readings under the same circumstance; I would also have worries if two different Beckmann thermometers gave me mutually inconsistent sets of readings. In addition, coherence with other things we believe also guides our thinking. If I am heating a body of water with a steady heat source, then it is reassuring to see that the mercury column in the Beckmann thermometer climbs up smoothly and steadily, not in a jerky fashion; conversely, if there are sudden changes in the situation, it is reassuring to see sudden changes in the thermometer readings. These are the kinds of factors that give us confidence that the new instrument actually constitutes an improvement, while it is not within the capacity of the old instrument to produce that judgement.

3.3. Handling recognized sources of error

One more story of temperature measurement will illustrate another dimension of progress in the absence of assured foundations. Daniel Gabriel Fahrenheit made some important early experimental contributions to the study of specific heats, by mixing measured-out amounts of different fluids at different initial temperatures and observing the temperature of the resulting mixture. In these experiments he was clearly aware of an important source of error: the initial temperature of the mixing vessel (and the thermometer itself) would have a distorting effect on the outcome. The only way to eliminate this source of error was to make sure that the mixing vessel *started out* at the temperature that the mixture would take, but that final temperature was unknown to Fahrenheit, since that was just what he was trying to find out by experiment. In a modern setting we would try to correct for this error by calculating the amount of heat exchange between the vessel and the liquids contained in it. But this would involve knowing the specific heat of the vessel and the liquids, and even the concept of specific heat was not available to Fahrenheit. Indeed, the concept could not have been established until after Fahrenheit's type of experiments started delivering stable results.

The solution adopted by Fahrenheit was both pragmatic and profound. In a letter of 12 December 1718 to Herman Boerhaave, he wrote:

(1) I used wide vessels which were made of the thinnest glass I could get. (2) I saw to it that these vessels were heated to approximately the same temperature as that which the liquids assumed when they were poured into them. (3) I had learned this approximate temperature from some tests performed in advance, and found that, if the vessel were not so approximately heated, it communicated some of its own temperature (warmer or colder) to the mixture.[11]

---

[11] P. van der Star (ed.), *Fahrenheit's Letters to Leibniz and Boerhaave* (Leiden: Museum Boerhaave; Amsterdam: Rodopi, 1983), 80–81.

I have not been able to find a record of the exact procedure of approximation that Fahrenheit used. However, the following reconstruction would be a possibility, and would be quite usable independently of whether Fahrenheit used it himself.

Start with the vessel at the halfway temperature between the initial temperatures of the hot and the cold liquids. Measure the temperature of the mixture in that experiment, and then set the vessel at that temperature for the next experiment, whose outcome will be slightly different from the first. This procedure could be repeated as many times as desired, to reduce the error arising from the initial vessel temperature as much as we want. In the end the initial vessel temperature we set will be nearly identical to the temperature of the mixture.

In this series of experiments, we knowingly start with an ill-founded guess for the outcome, but that guess serves as a starting point from which a very accurate result can eventually be reached. The initial affirmation of an existing system of knowledge may be made uncritically, but it can also be made while entertaining a reasonable suspicion that the affirmed system of knowledge is imperfect. The affirmation of some known system is the only option when there is no alternative that is clearly superior.

4. Progressive coherentism in action (II): chemical analysis

4.1. Elements and analysis

The second set of examples I would like to discuss comes from analytical chemistry.[12] Through much of the history of chemistry, firm knowledge about elements was something notoriously difficult to obtain. As chemists gradually inclined more toward the empirical and away from the metaphysical, it was perhaps inevitable that the question about chemical elements would be reformulated into a form amenable to empirical determinations. That is the origin of the analytical definition of chemical elements, which received its best-known formulation in Antoine-Laurent Lavoisier's work:

> ... if, by the term *elements*, we mean to express those simple and indivisible atoms of which matter is composed, it is extremely probable we know nothing at all about them; but, if we apply the

[12]   These examples were discussed at a presentation I gave at the workshop on 'Matters of Substance' at the University of Durham on 28 August 2006.

term *elements*, or *principle of bodies*, to express our idea of the last point which analysis is capable of reaching, we must admit, as elements, all the substances into which we are capable, by any means, to reduce bodies by decomposition. Not that we are entitled to affirm, that these substances we consider as simple may not be compounded of two, or even of a greater number of principles; but, since these principles cannot be separated, or rather since we have not hitherto discovered the means of separating them, they act with regard to us as simple substances, and we ought never to suppose them compounded until experiment and observation has proved them to be so.[13]

In Jean-Antoine Chaptal's more compact formulation, Lavoisier's definition amounted to taking as elements 'the endpoint of analysis'.[14]

This is often regarded as an operational definition, tying the concept of element down to the operational procedures of chemical analysis. As such, the analytical definition of element is seen as an important cornerstone of modern chemistry, as it left behind both the 4-element metaphysics of the Aristotelians and the 'philosophical' elements of the Paracelsians. However, the actual story is more complicated than that. In order to apply Lavoisier's definition, we need to know how to tell which chemical operations are decompositional. That requires us to know whether the products of a chemical operation are simpler than the ingredients going into it. But the whole point of the analytical definition of chemical element was to define simplicity in terms of decompositional operations, so we are caught in a tight circularity: what are simple substances?— those that cannot be decomposed further—what are decompositional operations?—those that produce simpler substances.

This circularity threatens to make the analytical definition of elements operationally empty. The difficulty can be illustrated easily through a few famous examples from the Chemical

---

[13] A. L. Lavoisier, *Elements of Chemistry, in a New Systematic Order, Containing all the Modern Discoveries*, R. Kerr (trans.) (New York: Dover, 1965; originally published in French in 1789, original English translation in 1790), xxiv; emphasis original.

[14] J. A. Chaptal, *Elemens de Chymie*, 3rd ed., 1796, vol. 1, p. 55; quoted in R. Siegfried and B. J. Dobbs, 'Composition, A Neglected Aspect of the Chemical Revolution', *Annals of Science* **24** (1968), 275–293, on 283.

Revolution.[15] Is sulfur an element? The phlogistonists didn't think so, because they thought (and showed by experiment) that it could be decomposed into sulfuric acid and phlogiston. The Lavoisierians disagreed, arguing instead that sulfuric acid was a composite substance, which they could decompose into sulfur and oxygen. Are metals elements? The phlogistonists didn't think so, as they knew how to decompose them into calxes and phlogiston. Again, the Lavoisierians saw the same reaction as a synthetic one, in which the metals combined with oxygen and formed oxides. Is water an element? Lavoisier said it was not, as he demonstrated by means of a hot gun barrel that removed the oxygen from water vapor and released the remaining hydrogen as a gas. Others made the decomposition even more clearly later by means of electricity, making a clean decomposition of water into hydrogen and oxygen gases. But Joahnn Ritter made the same experiment, and understood what he saw as the synthesis of hydrogen from water and negative electricity at the negative electrode, and oxygen from water and positive electricity at the positive electrode.

The analytical definition of element was in itself not sufficient to solve these debates. Something else was necessary for breaking the circle: either an independent method of judging the simplicity of substances, or an independent method of detecting decomposition when it took place, or both. How was this done? Much more work is required before I can answer that question in full historical detail, but I can offer a sketch that might be sufficient for the purpose of this paper. The trick was to turn the circle into a spiral, as it were. The analytical concept of chemical element developed step by step—building on, and improving, an uncertain foundation. Although the process of development was almost continuous, three rather distinct stages can be identified. The pattern of development we see in this case is the identification of a key regularity in a small subset of chemical reactions, which then gets codified as definitive for all chemical reactions, thereby changing the very definition of chemical element, and our view of what counts as chemical analysis and synthesis.

Stage 1. In the early days, chemists started with certain presumptions and untrained experience. It was presumed that

[15]   There is a large historical literature on the Chemical Revolution, but the best introduction to the subject for philosophers is A. Musgrave, 'Why Did Oxygen Supplant Phlogiston? Research Programmes in the Chemical Revolution', *Method and Appraisal in the Physical Sciences*, C. Howson (ed.) (Cambridge: Cambridge University Press, 1976), 181–209.

apparently homogeneous and durable substances were elementary (such as water and various metals). It was also presumed that certain harsh operations, such as the application of fire (combustion, distillation, melting) and treatment with acids, would result in the decomposition of various substances.[16] About elements, there were some metaphysical ideas that seemed to have rationalistic justifications. These included the idea that there must only be a small number of elements, and that there was an element ('principle') responsible for each distinctive general quality of matter (e.g., water as the principle of fluidity, and phlogiston as the principle of combustibility).

Stage 2. Out of the experience accumulated through the stage-1 conceptions and operations, a new view of chemical reactions emerged, which I will called the *component view*. According to the component view, chemical reactions were dissociations and re-associations of durable building-blocks; the simplest building-blocks were the elements. Ursula Klein locates the first articulation of this view in the work of Etienne-François Geoffroy and his fellow Paris academicians, reflecting on the successful laboratory operations performed by the metallurgists and pharmacists of the day.[17] The component view was initially founded on the reversibility of *some* chemical reactions, and the subsistence of chemical substances participating in those reactions (the ingredients of a synthetic chemical reaction maintaining their separate identity in the product, so that they could be recovered by analysis). Then the component view was extended and presumed to hold for *all* chemical reactions (including naturally occurring ones), expunging from chemistry the concepts of essence, maturation and enhancement, the contrast between active and passive substances, and animistic notions of affinity.[18] The comprehensive application of the component view disqualified certain presumed transformations (such as the alchemists' transmutation of elements) as chemical reactions, and altered the way many acknowledged chemical reactions were conceptualized. Thus, although the stage-1

[16] On the very long tradition of 'fire analysis' in alchemy and chemistry, and how it was later questioned by the likes of Robert Boyle, see A. G. Debus, 'Fire Analysis and the elements in the Sixteenth and the Seventeenth Centuries', *Annals of Science* 23 (1967), 127–147.

[17] U. Klein, 'Origin of the Concept of Chemical Compound', *Science in Context* 7, No. 2 (1994), 163–204, on 170 and *passim*.

[18] U. Klein, 'The Chemical Workshop Tradition and the Experimental Practice: Discontinuities within Continuities', *Science in Context* 9, No. 3 (1996), 251–287, on 256.

concept of chemical reaction was indispensable for generating the experimental knowledge that led to stage 2, it was partly negated and partly refined by the stage-2 concept of chemical reaction.

Stage 3. The stage-2 component view of chemical reactions led to the discovery of the conservation of weight in *some* chemical reactions (identified as such by stage-2 criteria). And then the conservation of weight was elevated to a universal principle in chemistry. This resulted in a focus on weight as the decisive indicator of the amount of substances, with some very important consequences. Most fundamentally, the focus on weight constituted an important refinement and change in the component view of chemical reactions, which had initially enabled the discovery of the conservation of weight. Weightless substances were eliminated from chemistry, even when they apparently maintained their identity through chemical combination and decomposition. Several accepted chemical compositions were reversed. For example, when phlogiston was eliminated from the conceptual arsenal of chemistry, sulfur and metals became elements and sulfuric acid and calxes became compounds.[19] Lavoisier's own theory suffered a change when later chemists rejected caloric as a substance; while Lavoisier regarded oxygen gas as a compound of 'oxygen base' and caloric, later Lavoisierians saw it as a possible state of the element oxygen. And, eventually, only reactions involving definite weight-ratios of substances were regarded as true chemical reactions, excluding physical mixtures such as alloys and solutions from the realm of chemistry. (Attempts to establish and explain the combining-weight regularities would soon lead to the development of the chemical atomic theory.)

4.2. Identification of substances

A similar story can be told about an even more basic task in chemistry: the phenomenological identification of substances (regardless of any ideas regarding their constitution). Here again, we have the kind of circularity that will sound familiar by now: the practical identification of a substance can only be made through the observation of its characteristic behavior, but how can we learn the

[19] These reversals are regarded as 'the real center' of the Chemical Revolution by Siegfried and Dobbs, op. cit. note 14, 281.

characteristic behavior of a substance, if we don't already know how to identify the substance as what it is? To put it more simply and abstractly:

(i)We have a question: 'Is this X?'
(ii)We answer it through our knowledge: 'X always does Y.'
(iii)But we cannot learn what X does, unless we can already identify X.

For example, we have a very old standard test for 'fixed air' (in modern terms, carbon dioxide), which is to bubble it through lime-water and check for the white precipitate.[20] But in order to discover that fixed air precipitated lime-water, chemists had to know in the first place how to identify fixed air (and lime-water, too).

The only possible answer here is that they had other criteria for identifying fixed air before the lime-water test was invented. But these criteria were neither precise nor infallible. The term 'fixed air' initially arose from the work of Stephen Hales (*Vegetable Staticks*, 1727), which showed that gases ('airs') could be 'fixed' in solid substances; any gas released from such solid combinations was also called fixed air.[21] It was Joseph Black's pioneering work published in 1755 on magnesia alba, quicklime and other alkaline substances that focused chemists' attention to the combination of what we now call carbon dioxide with solid substances, starting the process of turning 'fixed air' into a more precise category; Black also originated the lime-water test for fixed air.[22] With further

[20]   In modern terms, lime-water is an aqueous solution of calcium hydroxide ('slaked lime'), which reacts with carbon dioxide to produce calcium carbonate (chalk), which is insoluble in water so makes a precipitate. In chemical symbols, the reaction is: $Ca(OH)_2 + CO_2 \rightarrow CaCO_3 + H_2O$. It was known from early on that chalk ($CaCO_3$) could be made into caustic lime (CaO) by heating, but it was not recognized until Black's work that the process was a decomposition of chalk into lime and fixed air (CO_2). Caustic lime (CaO) becomes slaked lime ($Ca(OH)_2$) by absorbing water (H_2O), and slaked lime dissolves in water, making lime-water. See T. M. Lowry, *Historical Introduction to Chemistry*, rev. ed. (London: Macmillan, 1936), 61, and the rest of chapter 4.
[21]   For a brief introduction to the history of fixed air, see W. H. Brock, *The Fontana History of Chemistry* (London: Fontana Press, 1992), 78, 97–106, 124.
[22]   J. Black, *Experiments upon Magnesia Alba, Quick-Lime, and other Alcaline Substances*, Alembic Club Reprints, No. 1 (Edinburgh: William F. Clay, 1893), 24–25.

development of pneumatic chemistry in the hands of the likes of Joseph Priestley and Henry Cavendish, various kinds of gases were distinguished from each other by various properties. But 'fixed air' did not have a clearly distinctive property (such as oxygen supporting combustion and respiration) other than precipitating lime-water, so the lime-water test became quite definitive of fixed air. After these developments, some gases previously identified as fixed air were no longer considered to be fixed air; it is telling that Lavoisier had originally expected to generate fixed air in the very experiments that gave him oxygen. It was only after this kind of taxonomic tidying up that further and more precise properties such as the density, molecular weight and atomic composition of the gas could be determined precisely, leading to higher levels of theoretical sophistication. So, as in the other cases, knowledge started with bare sensations, theoretical presumptions and crude operations, and gradually built itself up with increasing precision, range, and coherence.

5. Epistemic iteration as the method of coherentist progress

The examples that I have discussed show different aspects of a particular mode of scientific progress, *epistemic iteration*, in which successive stages of knowledge are created, each building on the preceding one, in order to enhance the achievement of certain epistemic goals. In each step, the later stage is based on the earlier stage but cannot be deduced from it; the later stage often corrects results from the earlier stage as well as refine them. I conceive epistemic iteration in parallel with mathematical iteration, a computational technique in which a succession of approximations, each building on the one preceding, is used to achieve any desired degree of accuracy. But there are two important differences: mathematical iteration proceeds on the basis of a well-defined algorithm, which is usually lacking in epistemic iteration. Also, mathematical iteration has a pre-determined final destination, namely the rigorously correct solution, but that is generally not the case in epistemic iteration.

Epistemic iteration provides a key to understanding how knowledge can be improved without the aid of an indubitable foundation. What we have is a process in which we throw very imperfect ingredients together and manufacture something just a bit less imperfect. Niels Bohr once compared this epistemic process to dish-washing: we have dirty dishwater and dirty towels; still, somehow, we manage to get the dishes clean.[23]

The self-correcting character of epistemic iteration can be illustrated by a tale taken from everyday life (with a slight exaggeration). Without wearing my glasses, I cannot focus very well on small or faint things. So, if I pick up my glasses to examine them, I am unable to see the fine scratches and smudges on them. But if I *put on* those same glasses and look at myself in the mirror, I can see the details of the lenses quite well. In short, my glasses can show me their own defects. But how can I trust the image of defective glasses that is obtained through the very same defective glasses? In the first instance, my confidence comes from the apparent clarity and acuity of the image itself, regardless of how it was obtained. That gives me some reason to accept, provisionally, that the defects I see in my glasses somehow do not affect the quality of the image (even when the image is of those defects themselves). The next step, however, is to realize that some defects do distort the picture, sometimes recognizably. Once that is recognized, I can attempt to correct the distortions. For example, a large enough smudge in the central part of a lens will blur the whole picture seen through it, including the image of the smudge itself. So when I see a blurry smudge on my left lens in the mirror, I can infer that the boundaries of that smudge must actually be sharper than they seem. I could go on in that way, gradually correcting the image on the basis of certain features of the image itself that I observe. In that case my glasses would not only tell me their own defects, but also allow me to get an increasingly precise understanding of those defects.

I have only discussed a limited number and range of examples from real science in this paper. But it is easy to see epistemic iteration at work in many other areas and aspects of science. From some brief studies I have conducted, I am confident that measurement methods for various other physical quantities have also developed by iteration; good cases to study would include time, pressure, distance/length, weight, force, luminosity, and electric charge. It would be very interesting to see to what extent and in what ways theoretical developments are also iterative, as intimated by the brief look at early chemistry in the last section. And the examples considered so far suggest that a dynamic interaction between theory and observation will be a key feature in the iterative development of empirical knowledge.

---

[23]   See W. Heisenberg, *Physics and Beyond*, A. J. Pomerans (trans.) (London: George Allen & Unwin, 1971), 137.

Hasok Chang

Epistemic iteration is the central method of progressive coherentism. In the framework of coherentism, inquiry must proceed on the basis of an affirmation of some existing system of knowledge. This is precisely what happens in epistemic iteration. Starting from an existing system of knowledge means building on the achievements of some actual past group of intelligent beings. This gives the progress of knowledge an indelibly historical character and, ironically, also a conservative character.

At the same time, the conservatism of epistemic iteration is heavily tempered by an inherent pluralism. There are several aspects to this pluralism. We have freedom in choosing which existing system of knowledge to affirm as our starting point. The degree and depth of affirmation is also up to us. Perhaps most importantly, the affirmation of an existing system does not fix the direction of its development completely. The point is not merely that we do not know which direction of development is right, but that there may not even be such a thing as *the* correct or even the best direction of development. The desire to enhance different epistemic virtues may lead us in different directions of progress, since enhancing one virtue can come at the price of sacrificing another. There are a whole variety of important epistemic virtues that we may desire in a system of knowledge, such as consistency, precision, scope, simplicity, unity, explanatory power, intelligibility, fertility, testability, and so on.

We are often led away from a recognition of this inherent pluralism in science by an obsession with Truth. One might argue that the various epistemic virtues are not the correct measures of progress, but only Truth is. However, the difficulties of foundationalism force us to recognize that a reliable judgement of Truth, in the sense of correspondence to reality, is beyond our reach. Therefore, even if Truth should be the ultimate *aim* of scientific activity, it cannot serve as a usable *criterion* for evaluating systems of knowledge. If scientific progress is something we actually want to be able to assess, it cannot mean closer approach to the Truth.

All in all, the coherentist method of epistemic iteration points to a pluralistic traditionalism: each line of inquiry needs to take place within a tradition, but the researcher is ultimately not confined to the choice of one tradition, and each tradition can give rise to many competing lines of development. The methodology of epistemic iteration allows the flourishing of competing traditions, each of which can progress on its own basis without always needing to be judged in relation to others. Scientific progress requires tradition and allows freedom at the same time.

Action at a distance

ROBIN LE POIDEVIN

1. What is 'action at a distance'?

In the broadest sense of the phrase, there is action at a distance whenever there is a spatial or temporal gap (or both) between a cause and its effect. In this sense, it is not at all controversial that there is action at a distance. To cite a few instances: the page a few inches in front of you is impinging on your senses; the Sun is now warming the Earth; we are still living with the consequences of the Second World War. What *is* controversial is the idea of *unmediated* action at a distance, where there is both a gap between cause and effect and no intermediate causes and effects to fill it. The three examples just mentioned are cases of action at a distance, certainly, but not, surely, unmediated action at a distance. What we expect to find, in each case, is a spatially and temporally continuous causal series stretching across time and space.

Why has the possibility of unmediated action at a distance been taken seriously? According to the mechanistic world-view prevalent in the 17th century, and championed by Descartes, among others, the behaviour of matter was taken to be explicable entirely in terms of motion and collision. The paradigm instances of causal interaction (in the material world, at least) were therefore influence by contact, defined precisely by the absence of a gap. Newton's introduction of gravitational forces did not fit this rather rigid mechanical mould, and so appeared to illustrate instantaneous and unmediated causal influence across space. Newton himself made no such claim for gravitation, rejecting the idea of unmediated action at a distance as absurd. As he put it:

> That one body may act upon another at a Distance thro' a Vacuum, without the Mediation of anything else, by and through which their Action and Force may be conveyed from one to another, is to me so great and Absurdity, that I believe no Man who has in philosophical Matters a competent Faculty of thinking can ever fall into it. (Cohen (1978), 302–3)

21

Robin Le Poidevin

He nevertheless declined to postulate any mechanism for gravitation, famously declaring 'Hypotheses non fingo'.[1] Later, magnetic forces presented the same conundrum. The development of field theory in the nineteenth century showed how even gravitational and magnetic forces could be treated as mediated action at a distance.[2]

Unmediated action at a distance (hereafter simply 'action at a distance') is a violation of the *Principle of Locality*, the insistence that effects are local to their immediate cause, a principle supposedly threatened by certain implications of quantum physics. Perception of such a threat arose originally from reflection on a famous paper by Einstein, Podolsky and Rosen (1935), which raised the following problem. According to the 'standard' interpretation of quantum mechanics, the position and momentum of a particle are indeterminate until you measure them, and you cannot measure both position and momentum at the same time. But now consider a pair of particles, A and B, that are allowed to interact, and which then move apart. Suppose we then measure the momentum of one of the particles, A. Given information concerning the state of A and B during the period of their interaction, we can, from the value of A's momentum, infer B's momentum. *At that very moment*, then, B's momentum is determinate, even though we have not measured it directly, or interfered with it in any way. If instead we have chosen to measure A's position, we could have inferred B's position. So, if quantum mechanics is complete—if, in the words of the paper, 'every element of the physical reality must have a counterpart in the physical theory'—then both the momentum *and* position of B are determinate, independently of any direct measurement, contrary to the standard interpretation of quantum theory (Einstein et al. (1935), 779–80). Now, it is an assumption of this argument that measuring one particle cannot instantaneously affect the state of the other: that there is no instantaneous action at a spatial distance. But defenders of the standard interpretation may question that assumption: maybe measuring A does instantaneously affect B in a way not mediated by anything in the intervening space.

---

[1] 'I frame no hypotheses.' This remark occurs in the General Scholium Newton added to the 2nd edition of his *Principia*. A translation is provided in H.G. Alexander (1956), 164–71.

[2] For a historical survey of arguments concerning action at a distance, and its theoretical development, see Hesse (1961).

So what should the philosophical attitude to Locality be? Is the principle a necessary truth, a contingent but universal truth, or a largely true generalization with some admittedly bizarre exceptions? Perhaps, given its fragile status in contemporary physics, the proper attitude should be one of agnosticism. Here perhaps is yet another example of an entrenched view posing as an *a priori* truth, which must, like so many other metaphysical sayings before it, be surrendered to the fickle ways of scientific fortune. But perhaps it is possible to take a rather more robust position. Philosophers should certainly be concerned about the principle's status, since it is bound up with other important concepts. Let me give two examples.

The first example is persistence through time. What is it that underlies our ordinary judgement that a given object remains the *same object* through time? One influential answer to this is that it is *spatio-temporal continuity*: a life history should contain no spatial and temporal gaps.[3] Another answer is in terms of *causal continuity*: each stage in a life history should be intimately causally connected to its previous stages.[4] Locality shows that these are not actually competing criteria, for causal continuity implies spatio-temporal continuity.

The second example is the intrinsic/extrinsic property distinction. The intrinsic properties of a thing are often defined as the properties it has which do not logically (or perhaps, better, metaphysically) depend on the existence or properties of any other object. Being 100 cm^3 in volume, composed of carbon compounds, and having a temperature of 65° thus count as intrinsic. Being my niece's favourite pet, on top of the Matterhorn, and in the path of a beam of ultra-violet light do not. This characterisation has its limitations, however. Suppose substantivalism is true: space exists as an object in its own right, independently of its contents.[5] Then even being 100 cm^3 in volume depends metaphysically on a quite distinct object, namely a region of space. Indeed, any property of a spatial object would then count as extrinsic. Our intuitive grasp of intrinsicness is, in fact, a spatial one, and it seems better to build this into the definition: the intrinsic properties of an object are

---

[3] See, e.g., Locke's discussion of the persistence of organisms: Locke (1700), II, xxvii, 3–4.

[4] This is implicit in Locke's analysis of personal identity over time in terms of the continuity of memory: *op. cit.*, II, xxvii, 9.

[5] For a detailed characterisation and defence of substantivalism, see Nerlich (1994).

those that do not depend metaphysically on the existence or properties of anything outside the spatial boundaries of that object. That there is a genuine distinction to be made is confirmed by the thought that, if Locality is true, then the spatial conception of intrinsicness articulated above leads to a causal criterion:

A causally active property F is *intrinsic* to x if and only if the immediate effects of F are in x's immediate spatio-temporal vicinity

Of course, the criterion is inapplicable to acausal properties, which is why this is a described as a criterion, rather than an analysis of intrinsicness. But it does help to put the intrinsic/extrinsic distinction on a securer footing when we reflect that the properties that we intuitively pick out as intrinsic have local effects; those we pick out as extrinsic (typically) do not.

The fate of Locality, then, is philosophically significant. The purpose of this paper is not to assess Locality's compatibility with modern physics, but to explore various reasons, mainly of a philosophical kind, we might offer in favour of it, and how those reasons bear on our understanding of its status. So, for instance, some reasons might support the idea that Locality is a metaphysically necessary truth, others that is simply physically necessary. I am going to suggest that giving up Locality has some anomalous consequences for our understanding of causation (over and above the sheer oddness of action at a distance, of course). But this falls short of establishing Locality as a metaphysically necessary truth. The prospects for establishing the latter do not look good.

First, however, let us address the question of how precisely Locality should be characterised.

2. The Principle of Locality

A simple formulation of Locality is as follows:

(1) There is no spatial or temporal gap between a cause and its immediate effects.

An initial objection is that the notion of 'immediate' effects implies that the causal series exhibits a *discrete* ordering: for any given member of the causal series, there is a unique immediate successor. But what if the causal ordering is *dense*, such that between any two members of the series there is a third? Then we cannot, apparently, talk of the 'immediate effects' of a cause. The causal relation between any two members of a dense causal series will always be mediated.

We can, however, define 'immediate' in such a way that it is applicable to discrete and dense series alike, as follows:

For any cause c, a series (of events, states, facts or whatever) S contains the *immediate effects* of c if and only if S contains some effect x of c, and all causal intermediaries of x and c.

In a dense ordering, of course, S will contain no first member, but there is no effect of c closer than any member of S. (1) can therefore define Locality for both discrete and dense series. However, there is a further objection. Consider a dense causal series consisting of two parts, A and B, between which there is a spatial or temporal gap. Suppose further that A has no last member and B no first member. Any cause we choose in this series will satisfy (1): because of the peculiar topological properties of the series, there will be no gap between that cause and a set containing its immediate effects. However, the series clearly does not satisfy Locality, as we intuitively grasp it, since the series contains a spatial or temporal gap.[6]

An attempt to exclude this case is the following formulation:

(2) For any non-zero distance or interval d between any cause and effect in a series, there is an intermediate cause that is $d/2$ from the cause.

This condition would not be satisfied by the case we imagined. But it is not a satisfactory formulation. First (although this is perhaps a relatively minor worry), it presupposes that there is an objective division of an interval or distance into two equal halves. It presupposes, in other words, an objective space-time metric. But space and time, intuitively, might lack such a metric, and yet causal series not contain gaps. Second, even if we insist on such a metric, there is still the objection that Locality is a purely topological principle, not a metrical one. Third, suppose space-time to exhibit a discrete structure, with the consequence that causal series exhibit a discrete ordering. And let us suppose further that the distance between any two items is a function of the number of space-time 'atoms', or indivisible, partless minima between them. Now suppose there to be an odd number of such atoms between two items, adding up to distance d. Under these circumstances $d/2$ is not a possible location for any intermediate item. Fourth, if the

---

[6] I owe this objection, and formulation (2), to Timothy Williamson.

causal series describes a curved trajectory in space, and we conceive distance d between two items to be the shortest route between them, then there is no reason to suppose that there will be intermediate causes at $d/2$ (or indeed anywhere along that shortest route).

There is a further concern. As Lange (2002) has pointed out, spatio-temporal locality is not simply the conjunction of spatial locality and temporal locality. Consider the propagation of causal influence through space (a light sphere, for instance). Here, spatially intermediate causes should also be temporally intermediate. This is spatio-temporal locality. It would be possible for a sufficiently peculiar causal series to satisfy both spatial locality and temporal locality, and yet not to satisfy spatio-temporal locality. Here is his suggested formulation:

> For any event E, any finite temporal interval $\tau > 0$, and any finite distance $\delta > 0$, there is a complete set of causes of E such that for each event C in this set, there is a location at which it occurs that is separated by a distance no greater than δ from a location at which E occurs, *and* there is a moment at which C occurs *at the former location* that is separated by an interval no greater than τ from a moment at which E occurs *at the latter location*. (Lange (2002), p. 15)

Despite the reference to interval lengths in this characterization, there is no presupposition of an objective metric, because even if there is no fact of the matter as to whether a given spatio-temporal region is as large as any other non-overlapping region, it will still be true that a region will be larger than any region it contains as a proper part, which is all we need for Lange's definition. So we may justly regard it as a purely topological analysis. It is, however, subject to the same counterexample as (1), the causal series with an unusual topology, and the fourth objection to (2): the curved causal trajectory.

To summarise the discussion so far: we are looking for a non-metrical condition, one that is consistent with both dense and discrete causation, and which rules out *relevant* space-time gaps—i.e. ones that disrupt the causal series. The following promises to fit the bill without being over-complex:

(3) If x causes y, then either (i) x and y are contiguous in space-time, or (ii) x and y are linked by a causal chain that follows a continuous spatio-temporal pathway between x and y.

So much, then, for what Locality says. We turn now to the question of its status.

3. Locality as empirical hypothesis: Faraday's criteria for action at a spatial distance

Perhaps the Principle of Locality is no more than an empirical generalization, something which agrees with our ordinary experience of causation. If this is what it is, then we should be able to make sense of observations that would falsify it, evidence of action at a distance. What would constitute such evidence?

Empirical means of determining whether action over a spatial distance is mediated or unmediated are suggested in an 1852 paper by Faraday, entitled 'On the Physical Character of the Lines of Magnetic Force'.[7] His main concern in that paper is with the question 'whether the lines of magnetic force have a physical existence or not' (Faraday (1852), §3297). He compares and contrasts magnetic force with other kinds of force or causal influence: gravitation, radiation (of heat or light), electric current and induction. These other kinds of force seem to fall into three categories:

> Three great distinctions at least may be taken among these cases of the exertion of force at a distance: that of gravitation, where propagation of the force by physical lines through the intermediate space is supposed not to exist; that of radiation, where the propagation does exist, and where the propagating line or ray, once produced, has existence independent either of its source, or termination; and that of electricity, where the propagating process has intermediate existence, like a ray, but at the same time depends upon both extremities of the lines of force, or upon conditions (as in the connected voltaic pile) equivalent to such extremities. (Faraday (1852), §3251)

Gravitation he takes to be the paradigm case of (unmediated) action at a distance, although he is willing to entertain some doubt on the matter:

> There is one question in relation to gravity, which, if we could ascertain or touch it, would greatly enlighten us. It is, whether gravitation takes *time*. If it did, it would show undeniably that a

7 See Hesse (1961), 198–206.

physical agency existed in the course of the line of force. It seems equally impossible to prove or disprove this point; since there is no capability of suspending, changing, or annihilating the power (gravity), or annihilating the matter in which the power resides. (*Ibid.*, §3246)

The clearest case of mediated action over a distance is radiation. Here,

Lines of force have a physical existence independent, in a manner, of the body radiating, or of the body receiving the rays. They may be turned aside in their course ...The lines have no dependence upon a second or reacting body, as in gravitation, and they require time for their propagation. In all these things they are in marked contrast with the lines of gravitating force. (*Ibid.*, §3247)

The most important empirical criterion of action at a distance according to Faraday, then, is this:

(a)   the transmission of action is instantaneous.

In practice, however, it may be hard to establish whether this obtains or not. The rationale for this criterion is presumably this, that if the transmission of influence requires the existence of an intervening process, this will take time. A second criterion, although this is less prominent, is

(b)   the action depends on the simultaneous existence of the source and terminus, or reacting body.

The relevance of this criterion is less clear, but it may be a consequence of the time criterion. For if the transmission were mediated, then it could continue after the destruction of its source, and prior to the existence of a receiving body. Finally, a third criterion:

(c)   the direction of influence is unaffected by changes in the intervening space.

The rationale for this is obvious: if the transmission of influence require the existence of states in the intervening states, they will be susceptible to influences in that space.

Having considered in some detail the phenomena associated with magnetism, Faraday comes to the 'speculative' conclusion that, although the propagation of magnetic influence does not appear to take time, magnetic lines of force exist in the intermediate space between the objects concerned (for example, between two magnets whose north poles are facing each other).

How satisfactory are these as empirical criteria for action at a distance? Consider the key criterion, (a). The assumption is that spatial action at a distance is instantaneous. But if we are prepared to allow the existence of instantaneous causation across a spatial gap, it is unclear on what grounds we can insist that transmission of action through spatial intermediaries must take time. Why cannot every link be instantaneous? Perhaps, then, the suggestion is not that cause and effect are simultaneous in cases of spatial action at a distance, but simply that there should be no temporal gap between them. But this presupposes *temporal* Locality: there cannot be a temporal gap between cause and immediate effect. What is the justification for this? Why are space and time being treated differently in this respect?

This point also undermines criterion (b). For if violation of temporal Locality were to be allowed, it would no longer be clear why the effect should be co-existent with its source.

As for (c) we might have expected this to be augmented by the requirement that the effect be independent of the *distance* between the two reacting ends. True action at a distance ought not to be so dependent: since the causal influence does not in this case involve causes in the intervening space, it should be, as far as the strength of causal influence is concerned, as if there were *no* intervening space. Yet gravitational force *is* dependent on the distance between the ends, according to the Inverse Square Law. What explains this law, if the propagation of gravitational influence does not involve intervening causes? The answer has to be *nothing*: the Inverse Square law is simply a brute fact if gravitation is true action at a distance. If, however, gravitation is mediated, and we conceive of the field of gravitational influence as a sphere, then the Inverse Square Law follows as a geometrical consequence.[8] We can also deduce that, in a two-dimensional space, gravitational attraction would fall off as a simple inverse of the distance, and in four dimensional space, as the inverse of the cube of the distance.

Of course, it could be insisted that it is indeed just a brute law that the strength of interaction varies with the distance. But this undermines criterion (c). If variation with distance can be a brute law, why can variation with the medium not also be the subject of brute law? It seems, then, that all three of Faraday's criteria can be challenged. Are there other *a posteriori* means of determining

---

[8] See Lange (2002), 96.

whether a given phenomenon is a case of genuine action at a distance, or simply mediated? Faraday's criteria, note, are indirect. We might instead attempt to discover directly whether, in cases of a gap between cause and effect, there are intermediate causes. But insofar as this involves intervention, we cannot be sure that we have not introduced an intermediate cause (or effect) where there was none before.

This is hardly an exhaustive discussion of the empirical means of discovering action at a distance, but even this brief excursus on Faraday's criteria illustrates the difficulties involved in regarding Locality simply as an empirical, and so falsifiable, hypothesis. So, in the absence of unequivocal *a posteriori* methods, let us turn to *a priori* considerations.

4. Locality as metaphysical truth

Considered as a metaphysically necessary truth—one that *could not* be false, even if not analytically true—Locality poses a puzzle.

Suppose we grant that all causal chains from a particular source must, as a matter of necessity, proceed via effects that are local to the source. Then, in cases where there is both a spatial and a temporal gap between cause C and effect E, we have the following:

(i) There is an x that is causally intermediate between C and E, such that

(ii) x is spatially intermediate between C and E, and

(iii) x is temporally intermediate between C and E.

This is a putative example of something that Hume, for instance, thought impossible: a necessary connection between what he called 'distinct existences' (Hume (1739–40), Appendix, 635). By 'distinct existences' he meant logically distinct objects or states of affairs. That x is causally intermediate is logically distinct from its being temporally or spatially intermediate. Locality is not an analytic truth.

But Hume's conception of necessity is a narrow one: he conceives it as logical necessity. And that there could not be logically necessary connections between logically distinct existences is simply a trivial truth. But the last forty years or so has seen growing support for the notion of metaphysical necessity, where this is distinct from logical necessity. To take one of Saul Kripke's examples: any true identity statement, even if not analytic, such as 'Hesperus is Phosphorus', is necessarily true (Kripke (1972),

102–5). But where we find metaphysical necessities, we should expect to find items that are not ontologically distinct. In this case, the individual named by 'Hesperus' is not ontologically distinct from the individual named by 'Phosphorus': both name the planet Venus (and do so, if Kripke is right, in every possible world: *ibid.*, 102). So we can reframe Hume's injunction as follows: there are no metaphysically necessary connections between ontologically distinct existences. The puzzle, then, is this: if Locality is indeed a necessary truth, then the states of affairs represented by (i)–(iii) above should not be ontologically distinct, yet they appear to be so.

We could, of course, reject the metaphysical version of Hume's injunction. But that leaves us with a connection between causal and spatio-temporal continuity that is simply offered as a brute fact, inexplicable in terms of anything else, and inadequately motivated by empirical reasons. A more satisfying, though controversial, strategy suggests itself: take causal connections to be not entirely distinct from spatial and temporal connections. In other words, pursue a reductionist programme. But in what direction should the reduction go?

Suppose we take causal connections to be constructions from spatio-temporal ones. Then the fact that A and B are causally related is just the fact that they are spatially and temporally related in some way. This is precisely Hume's approach:

> We may define a cause to be 'An object precedent and contiguous to another, and where all the objects resembling the former are plac'd in like relations of precedency and contiguity to those objects, that resemble the latter.' (Hume (1739–40), 170)

Here Locality is built into the very characterization of the causal relation. Causal relations are, on this account, constructions out of the spatio-temporal relations of events. This is not a straightforward identification of causal relations with spatio-temporal ones, since the latter relations obtain between individual events, whereas the causal relation is said by Hume to obtain in virtue of a constant conjunction of event types.

Locality emerges from this as a metaphysical truth, though at the price of abandoning the very conception of causation that made Locality interesting. For what the Humean conception does is effectively to eliminate causation as a relation between individual events in favour of regularities. In contrast, a more robustly realist approach to causation treats those regularities as something that

Robin Le Poidevin

emerges from the individual causal relations.[9] According to this approach, it is because there is causation at the level of events that there are large-scale regularities, rather than vice versa. I will not attempt to defend this (surely intuitive) view here, but it is this view that invites a substantive answer to the question: why should the causal chains emanating from a given cause always proceed via the locality of that cause? It is worth considering whether a substantive answer can be given. Hume's analysis gives us the form of a response, but since we are now considering the alternative to the regularity account, let us reverse his analysis and propose that space and time are simply aspects of causation itself.

What would such an account look like? Take time first, as the more straightforward case. We can easily see how temporal precedence could be constituted by causation:

x occurs before y if and only if x is a cause of y

It follows from this that any causal intermediary between x and y will also be a temporal intermediary. Temporal Locality, however, requires more than this. The immediate effects of x must be in x's immediate temporal vicinity. Not all causal theories of temporal precedence will guarantee this. It is consistent with the above analysis that temporal separation is quite independent of causation. Suppose that time exists independently of the events that are located in it, and that times are ordered by betweenness relations. Such a series could exhibit order without any direction: a time series without an 'earlier than' relation. The direction of time could then be entirely a result of the directedness of causation. The temporal separation between x and y would then be purely a result of their location in time; but the fact that x is before y would depend on x's being a cause of y. In such a world, temporal Locality is not guaranteed. To guarantee temporal Locality, time would have to be completely reducible to causality.[10]

We may be able to reduce time to causality, but can we do the same with space? The connection between space and cause seems less intimate than that between time and cause. A number of objections, in fact, arise for a 'causal theory of space':

(a)   causal relations do not entail (non-zero) distance ones: what is happening at one time in the middle of the sun causes in part what happens in exactly the same place at a later time;

[9]   See, e.g., Tooley (1987).
[10]   For discussion of various versions of a causal theory of time order, and the problems each raise, see Le Poidevin (2003), Chapter 12.

(b)   distance relations do not entail causal ones: I can be 4000 light-years from a star without interacting with it;
(c)   space, unlike time and causation, is not intrinsically directed;
(d)   if we assent to a causal theory of time and space, we necessarily identify time with space.

To (a) we might respond that, although not all causal relations involve distance, some do: the propagation of light from a source, for instance. So perhaps space is a construction from certain kinds of causal processes. If this strategy is to work, however, it must be the case that a different kind of causal relation is involved in those cases that imply distance between cause and effect: the difference cannot simply lie in the *relata*. As for (b), if we accept the strong version of causal theory of time contemplated above, everything has to be causally connected in some way to everything else.

(c) and (d) are rather harder to deal with. An initially promising strategy is to suggest that causality is *multi-dimensional*. In fact, we have to say this if we are to combine the idea that space is reducible to causality whilst preserving the evident fact that space is multi-dimensional. The suggestion that causal relations exhibit more than one dimension is not immediately absurd. We talk, for instance, of the dimensions of sound, meaning simply that there are independent ways in which sounds may differ from each other: in pitch, timbre and volume. If causality, then, were four-dimensional, then we could identify one of these dimensions with time and the other three with space. Thus time would not be identified with space, and could exhibit a direction which the three dimensions of space lack.

The difficulty with this proposal is that, although it can explain both spatial Locality and temporal Locality, it cannot explain spatio-temporal Locality, as we defined it in § 2. In fact, although it can explain spatial Locality with respect to a single spatial dimension, it cannot explain why, where x and y are separated in more than one spatial dimension, any causal intermediary must be between x and y with respect to all those dimensions. The reason for this lies in the fact that variation in one dimension is, by definition, independent of variation in another dimension—this is what makes them different dimensions. If causality is multi-dimensional, then there is no reason to suppose that an ordering in one dimension will match an ordering in any other dimension.

What we have, then, is a rather complex and counter-intuitive conception of causality (more than one kind of causal relation, and at least one of these being multi-dimensional) that can explain only limited kinds of Locality.

The prospects, then, for a viable metaphysical picture of causation from which Locality emerges as an *a priori* consequence, seem bleak.

5. Locality as condition of physical law

Locality, we have suggested, is best represented neither as a mere empirical hypothesis, nor as a metaphysical truth. Could it, then, have an intermediate status?

Let us begin by posing this question: what do causes do? In general terms they determine the chances of their effects. If the world is deterministic, then (given background circumstances) they raise the chances of their effects to 1. If the world is indeterministic, then they simply raise the chances of their effects to something less than 1. But *where* and *when* those effects occur is not an arbitrary matter. What, then, determines the chances of the effects occurring at the time and place they do? Here are the possibilities:

  (i) cause alone;
 (ii) cause plus some feature of the location of the effects;
(iii) cause plus some further principle.

Option (i) is decidedly peculiar. The suggestion is that, in action at a distance, some intrinsic feature of the cause directly determines the gap between cause and effect. Let us suppose, for a moment, that the gravitational force that x exerts on y is an instance of this. According to (i), something intrinsic to x (and therefore logically independent of y) determines that the gravitational effects are felt precisely where and when y is. To see how unlikely this is, suppose further that y is being mechanically moved in an orbit around x by some device that is quite independent of x. The location of the gravitational effects of x is now constantly changing, and this, *ex hypothesi*, is explained by some corresponding change in the intrinsic properties of x. But of course, we know that x itself is not changing in this way, or that, if it is, any correspondence between those changes and the changes in y's location would be quite coincidental. For in the set-up imagined, the cause of y's location (namely, the mechanical device) is quite independent of the cause of the location of x's gravitational effects on y (namely, some intrinsic feature of x).

It is much more natural to suppose that the location of x's effects on y will be determined by y's location. After all, where else could x's effects on y be felt except where y is? Neither x nor y *causally* determine the location of those effects: the location of y is simply a logical constraint on where x's effects can be felt. This takes us to (ii). For gravitational effects to be felt by an object, y, there must another object, x. That is the causal condition. *Where* those effects are felt is logically determined by the spatial location of y. But what explains *when* those gravitational effects are felt? Our instinct here is to point to the temporal location of the *cause*. x's gravitational pull on y is felt at t because x exists at t. But such an appeal presupposes temporal locality: there can be no temporal gap between cause and effect. But if we allowing spatial action at a distance, temporal Locality looks rather less secure. Why should time alone obey Locality?

Since (ii) forces us to appeal to a principle, this response collapses into (iii): what determines the location of the effect is the (location of) the cause plus some other principle. What other principle? The natural candidate, surely, is the one that rules out *any* gap between cause and effect, namely Locality. This, as nineteenth century physicists discovered, has a serious ontological implication, namely the reality of fields. So when we ask why there are gravitational effects only *here*, where y is, the answer is that there are effects in other places too, but they are only manifested in certain ways where there is an object. The gravitational pull on the Earth is the result of the *local* presence of the gravitational field.

Without Locality, then, it would be completely mysterious why the effects of a cause occur at the time and place that they do. And unless there is in action a principle like Locality, making it non-accidental that effects occur when and where they do, there would be no *a priori* reason to suppose that there would be any regularities in nature. But such regularities are a necessary condition of physical law. The natural conclusion to draw is Locality is a condition of there being physical law.

Given that we live in a law-governed, non-chaotic world, that is a reason for thinking Locality is true. What it is not, however, is an *explanation* of the truth of Locality, for such an explanation would show how Locality followed from some more fundamental considerations. In the absence of such an explanation, Locality cannot reasonably be presented as a metaphysical truth. Its physical importance, however, is indisputable.

References

Alexander, H.G. (1956), *The Leibniz-Clarke Correspondence*, Manchester: Manchester University Press.

Cohen, I.B., ed. (1978) *Isaac Newton's Papers and Letters on Natural Philosophy*, Cambridge, Mass.: Harvard University Press.

Einstein, A., Podolski, B. and Rosen, N. (1935) 'Can Quantum-Mechanical Description of Physical Reality be Considered Complete?', *Physical Review* 47, 777–80.

Faraday, David (1852) 'On the Physical Character of the Magnetic Lines of Force', in *Experimental Researches in Electricity*, Vol. III, London: Richard Taylor and William Francis, 1855.

Hesse, Mary B. (1961) *Forces and Fields: the concept of Action at a Distance in the history of physics*, London: Thomas Nelson and Sons.

Hume, David (1739–40) *A Treatise of Human Nature*, ed. L.A. Selby-Bigge and P.H. Nidditch, Clarendon Press, 1978.

Kripke, Saul (1972) *Naming and Necessity*, Oxford: Blackwell.

Lange, Marc (2002) *An Introduction to the Philosophy of Physics: Locality, Fields, Energy, and Mass*, Malden, Mass.: Blackwell.

Le Poidevin, Robin (2003) *Travels in Four Dimensions: the Enigmas of Space and Time*, Oxford: Oxford University Press.

Locke, John (1700) *An Essay Concerning Human Understanding*, 4th edition, ed. P.H. Nidditch, Oxford: Clarendon Press, 1975.

Nerlich, Graham (1994) *The Shape of Space*, 2nd edition, Cambridge: Cambridge University Press.

Tooley, Michael (1987) *Causation: A Realist Approach*, Oxford: Oxford University Press.

Lessons from the History and Philosophy of Science regarding the Research Assessment Exercise

DONALD GILLIES

1. Introduction

The Research Assessment Exercise (henceforth abbreviated to RAE) was introduced in 1986 by Thatcher, and was continued by Blair. So it has now been running for 21 years. During this time, the rules governing the RAE have changed considerably, and the interval between successive RAEs has also varied. These changes are not of great importance as far as the argument of this paper is concerned. We will concentrate on the main features of the RAE which can be summarised as follows.

At intervals of a few years, RAEs are carried out in all the universities of the UK. The first step is to appoint a committee of assessors in each subject. These assessors are usually academics working in the field in question in the UK. Next most members of each department in a subject have to select a set of pieces of their research. The department then submits all these pieces of research produced by its members to the assessment committee. The members of the committee study this research output, and, on its basis, grade the department on a scale running form very good downwards. The departments which score well on the RAE are provided with research funds. Those which don't score so well are less fortunate. They are provided with much smaller funds for research, and the members of such departments have to spend more time on teaching. Recently there have even been moves in some universities to close altogether departments which perform badly on the RAE.

Such then in rough outline is the procedure followed in the RAE. It should be pointed out that the RAE is a costly operation, both in money terms and in the amount of academics' time which it absorbs and which they could, in the absence of the RAE, spend on their research. The question then naturally arises as to whether this expensive procedure has actually improved the research produced in the UK. Strange to say this question is rarely asked. Academics devote themselves with great energy to evaluating each other's work, but seem to be little concerned with evaluating an important government policy. Perhaps the reason for this is that it seems at first sight rather obvious that the RAE should improve the UK's research output. The procedure conforms to common sense. If we want to improve research, we should first find out who is doing good research and then give funding to the good researchers while withdrawing funding from the bad researchers. The RAE appears at first sight to be doing just this, and so the conclusion seems inevitable that introducing such a system will improve research output.

In social life, however, things are rarely simple, and judgements based on common sense can often mislead. The RAE, which is so costly in terms of money and time, is designed to improve the research output of the UK, but could it be having the opposite effect? Could it be making the research output of the UK worse instead of better? In this paper I want to argue that the RAE is indeed likely to have a negative effect on the research output of the UK. My argument will be mainly based, as the title suggests, on results which have been obtained from the study of the history and philosophy of science, and I will next consider why history and philosophy of science (henceforth HPS) is relevant to the RAE.

Let us begin with history of science. Historians of science study the great episodes of scientific advance, and the research programmes which led to exciting discoveries and to new and important knowledge. However, they also study the research programmes which failed to produce any advances, and the obstacles and difficulties which have sometimes stood in the way of scientific progress. All these matters are surely relevant to the design of a government policy intended to improve scientific research. Moreover it is not just the individual episodes which are relevant. One needs to analyse the underlying general principles which favour scientific advance, or, conversely, the general nature of the obstacles which impede scientific progress. This task of generalising from history of science falls to the practitioners of philosophy of science. Thus I think there can be little doubt that HPS is highly relevant to assessing the effectiveness of the RAE.

Of course, HPS is mainly relevant to the RAE as applied to science. For the purposes of this paper, I will take 'science' in a broad sense to include mathematics, computer science, and medicine as well as the standard natural sciences such as astronomy, physics, chemistry, etc. However, I will not include in 'science', the social sciences and the humanities. Although my focus then is on science in the sense just defined, and although by main arguments will be drawn from HPS, I will digress slightly in the next section by considering a branch of the humanities with which I am quite familiar—namely philosophy. Thus in the next section (section 2), I will raise the question of whether the RAE has improved philosophy in the UK. This digression will prove useful since it will suggest some methodological procedures which I will use when, in sections 3 to 7, I come to apply results from HPS to the problem of the effectiveness of the RAE.

2. Has the RAE improved philosophy in the UK?

I will begin by considering philosophy in the UK in the period 1900–1975, that is to say in the twentieth century but before the introduction of the RAE. This period of almost three quarters of a century was one of great brilliance as regards philosophy in the UK. It begins in the years before the First World War with the Cambridge school founded by Moore and Russell. Moore wrote an outstanding work on ethics while Russell made his remarkable contributions to logic and philosophy of mathematics. They were followed in the next generation by Keynes who was at this time still a philosopher rather than an economist, and who made important contributions to the philosophy of probability and induction. Wittgenstein also came to Cambridge at this time, and, working with Russell, he developed his early philosophy. After the First World War, philosophy continued to flourish at Cambridge. The prodigious Frank Ramsey died in 1930 at the age of only 26, but not before he made remarkable contributions to the philosophy of logic, probability and mathematics. In that same year, however, Wittgenstein obtained a post at Cambridge where he remained with short interruptions until 1947, rising to the position of professor. It was during this period that Wittgenstein developed his later philosophy, and wrote most of his book: *Philosophical Investigations*. Wittgenstein died in 1951, and *Philosophical Investigations* was published posthumously in 1953. Many people regard this book as the greatest philosophical masterpiece of the twentieth century.

After the Second World War, in the period 1945–75, philosophy continued to flourish in the UK, but Cambridge was no longer the main centre. Instead this period is dominated by the ordinary language school of philosophy at Oxford, and by Popper's school in London. Ordinary language philosophy had much in common with

Wittgenstein's later philosophy, but it was developed in a somewhat different way at Oxford by figures such as Ryle and Austin. Popper and his school focussed on philosophy of science. The school included not only Popper himself, but Lakatos, and, one might say, half Feyerabend. I say 'half Feyerabend' because Feyerabend started as Popper's assistant, and his first lectureship was in Bristol in the UK. However, he later worked in the USA and continental Europe. Still up to the death of Lakatos in 1974, Feyerabend visited London frequently giving regular lecture courses there.

The Oxford school and Popper's school were not on the best of terms. Their feud was foreshadowed by a famous argument between Popper and Wittgenstein which took place in 1946, and has become the subject of Edmonds and Eidinow's interesting 2001 book: *Wittgenstein's Poker*. However, a feud between two rival and very different schools of philosophy is surely yet another sign that philosophy was flourishing in the UK.

Moreover in the period 1945 to 1975, there were other significant developments in philosophy in the UK which lay somewhat outside the main schools just described. Perhaps the most important of these was the beginning of the philosophy of artificial intelligence. In his classic article published in *Mind* in 1950, Turing argued that computers could eventually equal if not surpass human beings in intellectual skills. Lucas replied in 1961 with a famous argument which used Gödel's incompleteness theorems to try to show that the human mind would always remain ahead of any possible computer.

So surveying philosophy in the UK in the years 1900–1975 one cannot but conclude that this was a brilliant period in which philosophy of the highest quality was produced in abundance in the UK. However, this was all before anyone had thought of introducing the RAE, and there was no system in operation which resembled the RAE in any way. Doesn't this begin to suggest that the RAE may be quite unnecessary?

At this point, however, a stern defender of the RAE might say: 'This is all very well, but there is never any good reason for being complacent. Even if things are going well, there is still always room for improvement, and the RAE was designed to achieve such improvement.' Well, the RAE may have been designed to bring about improvement, but did it succeed? The RAE has been with us since 1986. Has the philosophical output of the UK improved during these nearly 21 years?

Having formulated this last question, I have to confess that I am unable to answer it. I am genuinely unsure as to whether philosophy produced in the UK in the last 10 years say has been good or not. Naturally some pieces of UK philosophy have struck me as good, while others have appeared to me as quite bad. However, even these judgements I hold tentatively and with considerable uncertainty, while I am, at the same time, completely certain that others hold quite different opinions from mine about what is good and what is bad. Earlier I talked about philosophy in the period 1945 to 1975, and here I felt confident about the opinions I expressed. After the lapse of thirty years, a historical perspective has been obtained and it becomes easy to judge who were the important philosophers, and to evaluate the significance of their contributions. The situation is quite otherwise with contemporary philosophy where it is hard to say which philosophical works have a real importance, which are now esteemed only because of some passing fashion, and which are at present unjustly neglected because their true significance has not yet been grasped.

Just as I have been writing about philosophy in the UK in the period 1945 to 1975 which ended thirty years ago, let us imagine someone writing about the period 1986 to 2007 in the year 2037. Can we guess what such a writer might say? There could be a spectrum of judgement along a scale running from those most favourable to the RAE to those most hostile to it. A very favourable judgement might go like this: 'Although few realised it at the time, the year 1986 proved a turning point for philosophy in the UK. The introduction of the RAE in that year stimulated the philosophy community in the UK to greater efforts and we can now recognize that the works they produced in the subsequent decades outshone those completed in the first three quarters of the twentieth century.' A very hostile judgement might go like this: 'There was a brilliant flowering of philosophy in the UK in the first three quarters of the twentieth century. Unfortunately after about 1975, a decline set in and the situation was made worse by the introduction of the RAE in 1986. In the decades following this introduction, philosophy in the UK, which earlier in the century had been so brilliant, sank to a very low point.' These two judgements could be taken as marking the extreme points of a scale. Where along this scale will the majority of writers in 2037 be found? I really would not like to say.

One important point has emerged from the preceding discussion, namely that it is often very difficult to judge the quality of contemporary research. It is often only after the elapse of a considerable interval of time—say thirty years or more—that one can say with any confidence that a piece of research was either genuinely good or really very bad. The situation is indeed worse than this, for, as we shall see later in the paper, contemporary judgements on the quality of pieces of research, are often wildly mistaken in the light of what emerges later on. But this exposes a key weakness in the RAE. The RAE naturally relies on contemporary judgements as to which researchers are good and which are bad, but such judgements are difficult to make and may often be found to be quite wrong in the light of later developments.

I will pick up this point again in what follows, but I would like now to consider another approach to assessing whether the effect of the RAE on philosophical output is likely to be good or bad. As we have seen the direct approach of asking whether the introduction of the RAE improved the quality of philosophy in the UK did not yield any very clear answer. This is why I suggest using another approach which I call the 'counterfactual methodology'. The idea here is to consider research carried out in the past, and ask whether, if the RAE had been in existence at that time, it would have improved that research or would, on the contrary, have made it worse.

Let us apply this counterfactual methodology to philosophy in the UK in the 1930s. Suppose the RAE had been introduced in say 1936, what effect would it have had on philosophy in the UK. This was the time when Wittgenstein was developing his later philosophy at Cambridge and writing early drafts of what later became his *Philosophical Investigations*, judged by many to be the greatest philosophical book of the twentieth century. How would Wittgenstein have fared in a RAE conducted according to the existing rules?

Actually we can answer this question quite easily. Wittgenstein was offered a position at Cambridge in 1930, rose to becoming professor there, and resigned from his chair in 1947. During these 17 years he published nothing. In fact the last philosophical work which he published in his lifetime was a paper entitled: 'Some Remarks on Logical Form' which appeared in the Journal of the Aristotelian Society in 1929. Wittgenstein had agreed to give a talk to the Aristotelian Society that year. The Aristotelian Society insists that papers are always printed in advance, and this is why Wittgenstein's paper was published. However Wittgenstein decided shortly after the printing had taken place that the paper was worthless, and at the meeting, actually talked on another topic (see Monk, 1990, 272–3 for details). After this experience, Wittgenstein became very reluctant to publish anything which he had not worked on for a long period, and this explains why he published nothing further for the next 17 years.

Now what happens under the RAE rules to academics who publish nothing? They are classified as research inactive, and their fate is not agreeable. Their research time is removed, and they have to spend more time teaching. Moreover they are at risk of being sacked. If the RAE had been in existence in 1936, and the rules had been applied without fear or favour, then this is the fate which would have overtaken Wittgenstein.

Now a defender of the RAE might at this point object to my analysis on the following grounds: 'Wittgenstein published nothing between 1930 and 1947 because he was under no pressure to do so. Had the RAE been introduced in 1936, he would certainly have 'knuckled under' and published some stuff.' Unfortunately for this argument, numerous memoirs and the magnificent (1990) biography of Wittgenstein by Ray Monk have given us quite a vivid picture Wittgenstein's character, and this leaves no doubt that Wittgenstein was the last person on earth to have 'knuckled under' and obeyed the directives of the RAE.

In fact Wittgenstein, despite his great intellectual brilliance, seems to have disliked the company and habits of academics and to have preferred associating with simple folk. Karl Britton, a former student of Wittgenstein's, very clearly describes this attitude of the master (quoted from Pitcher, 1964, 12):

> He had, he said, only once been to high table at Trinity and the clever conversation of the dons had so horrified him that he had come out with both hands over his ears. The dons talked like that only to score: they did not even enjoy doing it. He said his own bedmaker's conversation, about the private lives of her previous gentlemen and about her own family, was far preferable: at least he could understand why she talked that way and could believe that she enjoyed it.

As a result of these attitudes, Wittgenstein showed a strong propensity to abandon seats of academic learning, and go off to remote spots in the country where he could associate with simple country folk. This propensity manifested itself as early as 1913, where he decided to go off to live alone in a remote area of Norway for two years. Russell tried in vain to dissuade him, and wrote about it in a letter as follows (quoted from Monk, 1990, 91):

> I said it would be dark, & he said he hated daylight. I said it would be lonely, & he said he prostituted his mind talking to intelligent people. I said he was mad & he said God preserve him from sanity. (God certainly will.)

This episode illustrates the extreme obstinacy and determination of Wittgenstein's character. He was really not the sort of man who would have been prepared to 'knuckle under' and obey some government regulation which he regarded as mistaken.

Wittgenstein went to Norway but did not stay for two years because of the outbreak of the First World War. After the War, despite having become famous in philosophical circles because of the publication of his *Tractatus*, he decided to give up philosophy and worked as a schoolmaster in remote Austrian villages between 1920 and 1926. He refused to attend any of the meetings of the Vienna Circle which greatly admired his work. However Wittgenstein was eventually persuaded to return to academic life in Cambridge in 1930. Yet he remained full of longings for a simple life of manual toil in some remote country location. He even applied in 1935, at the height of Stalinism, to work as a labourer on a collective farm in Russia. Perhaps luckily for him the Russians turned down his application (see Monk, 1990, 351). So Wittgenstein went back to his hut in Norway for a year instead.

Many people might regard the job of being professor of philosophy at Cambridge as rather an agreeable one, but not so Wittgenstein. In a letter to Malcolm in 1945, Wittgenstein wrote: ' ... the absurd job of a prof. of philosophy ... is a kind of living death' (quoted from Malcolm, 1958, 38). At this time he was contemplating resigning his professorship at Cambridge, which he did in 1947. Just before his resignation, Wittgenstein wrote (quoted from Monk, 1990, 516):

> Cambridge grows more and more hateful to me. The disintegrating and putrefying English civilization. A country in which politics alternates between an evil purpose and *no* purpose.

After resigning his chair, Wittgenstein went off in 1948 to live in a remote country district in Galway on the west coast of Ireland.

These episodes give a vivid illustration of Wittgenstein's character and tastes. In the light of these, is it possible that, if the RAE had been introduced in 1936, he would have agreed to its demands and started publishing some of his work? I find it quite inconceivable that he would have done so. Malcolm in his Memoir records (49) that in the academic year 1946–7, Wittgenstein stated that 'he was not going to be 'stampeded' into publishing prematurely.' In fact he had published nothing for over 17 years at that stage of his career.

If there is still any doubt on this point, it could be added that Wittgenstein was also highly contemptuous of the typical academic procedures which are enshrined in the RAE. This is illustrated vividly in letters written by Wittgenstein to Malcolm in 1945 and 1948. It should be explained that Wittgenstein was very fond of reading American detective magazines—particularly those published by Street and Smith. In the 1930s and 1940s, *Mind* was a leading English philosophy journal, as indeed it still is today. To have a series of papers published in *Mind* would be regarded as a strong point in favour of any researcher according to the usual RAE criteria. Wittgenstein, however, far from endorsing these RAE criteria is very sarcastic about them, and compares *Mind* unfavourably with the detective magazines of Street and Smith (cf. Malcolm, 1958, 32). In 1945 he wrote to Malcolm:

> If I read your mags I often wonder how anyone can read 'Mind' with all its impotence & bankruptcy when they could read Street & Smyth mags. Well, everyone to his taste.

In another letter to Malcolm in 1948, he elaborated the comparison:

> Your mags are wonderful. How people can read Mind if they could read Street & Smith beats me. If philosophy has anything to do with wisdom there's certainly not a grain of that in Mind, & quite often a grain in the detective stories.

Suppose then that the RAE had been introduced in 1936. Are we seriously to suppose that Wittgenstein would have 'knuckled under' and submitted papers for publication in Mind? Given his character and views, it is altogether out of the question that he would have done so. His reaction is entirely predictable. In the face of such a demand, he would have undoubtedly have left Cambridge in disgust and gone off to his hut in Norway.

The effect of the introduction of the RAE in 1936 would then have been to hound Wittgenstein out of Cambridge. Hardly a result which should increase our confidence in the merits of the RAE! What actually happened was that Wittgenstein was offered a Chair in Philosophy at Cambridge in 1939, despite having published nothing for ten years. Such an appointment would of course be almost impossible under a RAE regime. Even if the members of the appointments panel were sympathetic to a candidate who had

Donald Gillies

published nothing for ten years, they could hardly overlook the fact that such a professor would contribute nothing to the RAE, and would indeed set a bad example to the rest of the department. So if the RAE had been introduced in 1936, Wittgenstein would have been very unlikely to have become Professor of Philosophy at Cambridge.

Wittgenstein was perhaps rather excessively reluctant to publish, but how can we condemn his strategy in general terms? Wittgenstein was not of course really research inactive while at Cambridge. Although he published nothing between 1929 and 1951, he produced roughly thirty thousand pages of notebooks, manuscripts, and typescripts on philosophy in that period (Malcolm, 1958, 84). That is an average rate of about 26 pages a week. Wittgenstein's view was that he shouldn't publish anything until he had thought and rethought about it, and worked through it many times revising and correcting. He believed that only in this way could he produce philosophical work of lasting value. Now how can we say he was wrong about this? After all, his strategy worked. At the end of his long years of rethinking, revising and correcting, he produced a book (*Philosophical Investigations*) which many regard as the philosophical masterpiece of the twentieth century.

I am not saying that every philosopher should adopt Wittgenstein's strategy. Other philosophers work in a quite different way and yet produce just as good philosophy. It is partly a matter of style and temperament. Russell, for example, who was in my opinion just as good a philosopher as Wittgenstein, worked in quite a different way. He had no inhibitions about publishing, and, when thinking about a problem, would often publish in rapid succession a series of papers considering different solutions before finally settling on a particular approach. But, although Wittgenstein's way of working is not the only one, it is certainly a possible way of working which has produced great philosophy. It is thus obviously wrong for the RAE to rule out this strategy of delaying publication, and this is a great weakness of the whole system.

Let us now consider what response a defender of the RAE might give to this objection. He or she might reply along the following lines: 'Philosophy is a peculiar intellectual discipline, and tends to attract peculiar people. Even by the standards of philosophers, Wittgenstein was exceptionally strange. Now if we turn from philosophy to more serious intellectual disciplines such as mathematics, medicine, physics or astronomy, we shall find that these scientific disciplines are carried out by more serious people for whom the criteria of the RAE are certainly appropriate.' To meet this challenge, we must turn to a consideration of science, and this therefore is a good point at which to begin the main line of argument of the paper.

In fact we will discover that many of the great scientific innovators had personalities which were no less unusual than Wittgenstein's. It will also emerge that Wittgenstein had some advantages which several of those who made great advances in science lacked. Wittgenstein's work was recognised very early on by individuals such as Russell and Keynes who could exercise a powerful influence in the academic world. Some other notable pioneers had the less agreeable experience of finding that their innovative work was not recognised by anyone, and indeed was rejected as absurd by those in powerful academic positions.

In considering how the RAE might affect research in science in the UK, I will apply the 'counterfactual methodology' introduced by the case of Wittgenstein. I will consider a number of great advances in science which occurred in the past, and ask whether, if the RAE had existed in those days, it would have helped or hindered that advance. The result of the cases which I will consider is the same as the result in the case of Wittgenstein—namely that the RAE, if it had been in existence, would constituted an obstacle to the advance.

I have chosen three cases which are designed to cover a range of different sciences. The first is in mathematics, the second in medicine, and the third in astronomy. I have chosen cases where a very striking scientific advance was made at a theoretical level. I do not, however, want to focus on theory and neglect practical applications. It is now generally agreed that the development of new technological applications of science is very important in order to make the UK competitive in the era of globalisation. I have therefore chosen three theoretical advances which had very important wealth-generating applications.

3. First Case-History: Frege and Mathematical Logic

My first example is taken from the field of mathematics and I want to consider an important advance made in a branch of the subject known as mathematical logic. This advance was made by Frege in a booklet published in 1879, and which is usually referred to by its German title of *Begriffsschrift*, which means literally: 'concept-writing'. It might be objected to this example that Frege was a

Donald Gillies

philosopher rather than a mathematician. It is true that Frege wrote some very important works on philosophy, but that does not make him any less a mathematician. Other famous mathematicians such as Descartes and Leibniz also wrote on philosophy. Frege worked all his life in the mathematics department of Jena university. The *Begriffsschrift* does contain some interesting philosophical remarks, but it is mainly formal in character. Its contribution is to what is now called mathematical logic and it is difficult to deny that mathematical logic is a branch of mathematics.

Indeed Frege's *Begriffsschrift* may justly be said to have introduced modern mathematical logic. In this work Frege presents for the first time an axiomatic-deductive development of the propositional calculus and of the predicate calculus (or quantification theory). The propositional and predicate calculi are the first things introduced in any modern treatment of mathematical logic. What is still more surprising is that the expositions of these calculi in contemporary textbooks are often quite close to the original expositions of Frege. Two well-known and widely used textbooks of mathematical logic are Mendelson (1964) and Bell and Machover (1977). Mendelson introduces the propositional calculus and quantification theory in chapters 1 & 2, while Bell and Machover introduce them in chapters 1, 2 & 3. Of course they both give many results and approaches which were discovered after Frege, but they do also give an axiomatic-deductive treatment which has a lot in common with Frege's and indeed uses some of the same axioms that Frege used.[1] Frege's treatment in the *Begriffsschrift* includes what is known as higher-order logic, whereas modern treatments usually limit themselves to first-order logic. However, leaving this subtlety aside we can say that Frege's treatment of both the propositional and predicate calculi is complete from a modern point of view, though his axiomatic presentation was subsequently simplified by reducing the number of axioms. Thus Frege created in the *Begriffsschrift* a whole new formal theory which is still today taken as the core of mathematical logic.

Frege's remarkable achievement has been fully recognised by experts in the field since the 1950s. In Appendix II to his English translation of the *Begriffsschrift*, Bynum very usefully collects together some evaluations by well-known scholars writing in the

---

[1] A detailed comparison of the *Begriffsschrift* with the treatment of the corresponding material in Mendelson (1964) and Bell and Machover (1977) is to be found in Gillies, 1992, 275–6.

Lessons from the History and Philosophy of Science

1950s and 1960s. Here are some extracts from the passages he gives. They are all quoted from Bynum, 1972, 236–8.

> Quine, 1952 (236): ' ... the logical renaissance might be identified with the publication of Frege's *Begriffsschrift* in 1879 ... 1879 did indeed usher in a renaissance, bringing quantification theory and therewith the most powerful and most characteristic instrument of modern logic ... with the aid of quantification theory modern logicians have been able to illuminate the mechanism of deduction in general, and the foundations of mathematics in particular, to a degree hitherto undreamed of.'

> Dummett, 1959 (238): 'There can be no doubt that Boole deserves great credit for what he achieved ... however ... Boole cannot correctly be called "the father of modern logic". *The discoveries which separate modern logic from its precursors are of course the use of quantifiers ... and a concept of a formal system, both due to Frege and neither present even in embryo in the work of Boole.*'

> Bochenski, 1962 (237): 'Among all these logicians, Gottlob Frege holds a unique place. His *Begriffsschrift* can only be compared with one other work in the whole history of logic, the *Prior Analytics* of Aristotle. The two cannot quite be put on a level, for Aristotle was the very founder of logic, while Frege could as a result only develop it. But there is a great likeness between these two gifted works.'

> William and Martha Kneale, 1962 (236–7): 'Frege's *Begriffsschrift* is the first really comprehensive system of formal logic... . Frege's work ... contains all the essentials of modern logic, and it is not unfair either to his predecessors or to his successors to say that 1879 is the most important date in the history of the subject.'

Frege carried out his researches in mathematical logic for purely theoretical reasons, but, as so often happens, his results turned out to be of great practical importance. Mathematical logic is one of the fundamental tools of present-day computer science, and one can further say that the computer as we know it today could not have developed without a prior development of mathematical logic. Detailed accounts of the use of mathematical logic in computer science and in the development of computing are contained in Davis (1988a & b) and in Gillies (2002). There are many specific examples of the application of mathematical logic in computer science, but at a very fundamental level one can say that the *Begriffsschrift* is the first example of a fully formalised language, and so, in a sense, the precursor of all programming languages (see Davis, 1988b, 316).

Thus Frege's research turned out to provide some of the fundamental tools for a wealth-generating technological advance. Consequently Frege's research work must be the kind of research work which a nation like the UK should try to encourage. This brings us to the question of whether Frege's research would have been helped if, counterfactually, there had been a RAE regime operating in Germany in his day. Suppose there had been a German RAE in the 1880s, how would Frege have done? The answer is: 'not very well.'

In Appendix I to his translation of the *Begriffsschrift*, Bynum gives in full the contemporary reviews of the work, all written in the years 1879 and 1880. It is very interesting to compare these with the evaluations of the same work made with the benefit of historical perspective in the 1950s and the 1960s. These are given by Bynum in his Appendix II, and we have already quoted some passages.

Turning now to the contemporary reviews of the *Begriffsschrift*, they were 6 in number—all quotations from them will be from the versions in Bynum, 1972, 209–35. Four were written by Germans, one by a Frenchman (Tannery) and one by an Englishman (Venn). Only one of these reviews, which was written by a German: Lasswitz, is favourable. The other three German reviews do make some favourable remarks, but one cannot help wondering whether these are designed to be polite to a compatriot and colleague, since they are contradicted by the majority of the detailed comments on the work which are highly unfavourable. Thus Hoppe concludes his review by saying (210): 'On the whole, the book, as suggestive and pioneering, is worth while.' However earlier in the same review he had written (209): '... we doubt that anything has been gained by the invented formula language itself.' Similarly Michaelis concludes his review (218): 'His work ... certainly does not lack importance.' However, this rather contradicts the following harsh judgement given in the body of the review, where Michaelis says (217): '... Frege has to pass over many things in formal logic and detract even more from its content... . The content of logic which has been much too meagre up to now, should not be decreased, but increased.' In contrast to the later critics who saw Frege as having made an enormous step forward in logic, Michaelis actually thinks that Frege has decreased, or detracted from, the content of logic.

However, the harshest German review comes from the most famous German logician of the time: Schröder. Schröder actually upbraids Lasswitz for having written a review supporting the *Begriffsschrift*, and says of this review (220) that he casts 'a disapproving glance at it'. He refers to Lasswitz later (221) as 'the Jena reviewer', which seems to imply that Lasswitz's favourable judgement arises from some personal connection with Frege. Schröder's own judgement on the *Begriffsschrift* is very negative indeed. He thinks that Frege has done nothing which has not already been done much better by other people. As he says (220): '... the present little book makes an advance which I should consider very creditable, if a large part of what it attempts had not already been accomplished by someone else, and indeed (as I shall prove) in a doubtlessly more adequate fashion.' It soon becomes clear that this other person is Boole. Indeed Schröder goes on to say (221) that, leaving aside the question of function and generality and some applications, '... the book is devoted to the establishment of a formula language, which essentially coincides with Boole's mode of presenting *judgements* and Boole's calculus of judgments, and which certainly in no way achieves more.' Here Schröder does seem to make an exception in favour of Frege's treatment of generality but this appearance is deceptive for he later goes on to say that Frege's treatment of generality is in no way superior to the Boolean. He writes (229–30): 'Now in the section concerning "generality", Frege correctly lays down stipulations that permit him to express such judgements precisely. I shall not follow him slavishly here; but on the contrary, show that one may not perchance find a justification here for his other deviations from Boole's notation, and the analogous modification or extension can easily be achieved in Boolean notation as well.' (Logicians will at once see from this that Schröder has completely failed to grasp the importance of introducing the quantifiers.) But could Frege at least be defended on the grounds that he has shed some light on the logical nature of *arithmetical* judgements? 'Not so', argues Schröder, 'for that matter too has already been cleared up by someone else.' In his own words (231): 'According to the author, he undertook the entire work with the intention of obtaining complete clarity with regard to the logical nature of *arithmetical* judgements, and above all to test "how far one could get in arithmetic by means of logical deductions alone". If I have properly understood what the author wishes to do, then this point would also be, in large measure, already settled—namely, through the perceptive investigations of Hermann Grassmann.' After dismissing Frege's work so completely, it is rather surprising that Schröder concludes (231): 'May my comments, however, have the over-all effect of encouraging the author to further his research, rather than discouraging him.' Perhaps Schröder felt some pangs of guilt about writing so harshly about the work of a young researcher in his field. The two non-German reviews of the *Begriffsschrift* are if anything even more dismissive than the German reviews, and contain no favourable remarks at all. Tannery in France writes (233): 'In such circumstances, we should have a right to demand complete clarity or a great simplification of formulas or important results. But much to the contrary, the explanations are insufficient, the notations are excessively complex; and as far as applications are concerned, they remain only promises.' Nowadays one of Frege's great advances is considered to be the replacement of the Aristotelian analysis in terms of subject and predicate by an analysis using function and argument. Tannery notes this change but regards it as a mistake (233): 'The [author] abolishes the concepts of *subject* and *predicate* and replaces them by others which he calls *function* and *argument....* . We cannot deny that this conception does not seem to be very fruitful.' Finally Venn in England entirely agrees with Schröder that Frege has made no advance over Boole and has indeed taken a step backwards. Venn writes (234): '... it does not seem to me that Dr. Frege's scheme can for a moment compare with that of Boole. I should suppose, from his making no reference whatever to the latter, that he has not seen it, nor any of the modifications of it with which we are familiar here. Certainly the merits which he claims as novel for his own method are common to every symbolic method.' Venn, moreover, has no kind words at the end of his review, but concludes by saying (235): '... Dr Frege's system ... seems to me cumbrous and inconvenient.' It is worth noting here that Frege's advances over Boole which seem so obvious today and which are mentioned by Dummett in the passage quoted above, were not appreciated at all by Schröder and Venn—two of the leading logicians of Frege's time. So to sum up. If we go carefully through the six contemporary reviews of Frege's *Begriffsschrift*, we find only one which takes a positive view of Frege's work. In the other five, there is a consensus to be found that the *Begriffsschrift* makes no advance on what has already been done, particularly by Boole and the Booleans, and indeed that it is in many respects inferior to and a step back from already existing logical works.

What is remarkable is that Frege was not discouraged by these damning reviews, but continued his work on his logicist programme for the next 24 years. However, his subsequent books were, if anything, even less successful than the *Begriffsschrift*. The *Foundations of Arithmetic* published in 1884 received only 3 reviews—all unfavourable. In 1891 Frege wanted to publish a third book in the series, but, perhaps not surprisingly, found it hard to find a publisher. Eventually, however, (Bynum, 1972, 34): 'the publisher Hermann Pohle in Jena ... agreed to print the book in two instalments, the publication of the second part to be dependent upon a good reception of the first. So, in late 1893, the first volume of *The Basic Laws of Arithmetic* appeared.' This book got only two reviews—both unfavourable. In the light of this Frege had to publish the second volume which appeared in 1903, at his own expense. In the 1890s and 1900s a few avant-garde researchers— notably Peano and Russell—did begin to study and develop Frege's ideas. However, even when Frege retired from Jena at the age of 70 in 1918, general recognition had still eluded him. The situation was well summed-up by Bochenski in 1962 (quoted from Bynum, 1972, 237–8):

> It is a remarkable fact that this logician of them all had to wait twenty years before he was at all noticed, and another twenty before his full strictness of procedure was resumed by Lukasiewicz. In this last respect, everything published between 1879 and 1921 fell below the standard of Frege, and it is seldom attained even today.

Anyone who is concerned with formulating policies concerned with research, should in my opinion read carefully the two appendices to Bynum, 1972. They amount to only 30 pages, but they demonstrate in a conclusive fashion that the method of peer review can, in some cases, go very wrong. It does happen that the majority of contemporary researchers in a field can judge as worthless a piece of research which is later, with the benefit of historical perspective, seen as constituting a major advance.

Now the RAE does clearly rely on peer review because the value of each researcher is judged by a committee of experts in the field. Indeed the RAE in a sense involves a double use of peer review, because the members of the RAE consider only research which has been published, and, to get a piece of work published, a researcher has usually to submit it to a journal which uses peer review to assess whether it is worth publishing. The problem facing those like Frege, whose work is judged of little value by the majority of their peers, is that they may find it difficult to publish at all. This applied to Frege himself. He wrote two papers replying to the criticism of Schröder and Venn that his work was inferior to that of Boole.

However he was unable to publish these papers (cf. Bynum, 1972, 21), and they only appeared long after Frege's death. Those who find their peers against their work will certainly be excluded from publishing in the more famous journals and may have to resort to publishing in lower quality journals or even to publishing the material in book form at their own expense, as Frege did in 1903. Now theoretically the RAE committee reads carefully and judges on their merits all the works submitted to it, but of course in practice papers which have appeared in high ranking journals, or books which have been published by prestigious firms such as Oxford University Press, are likely to be judged more favourably. Conversely papers which have appeared in low ranking journals, or, worse still, books published at the author's own expense—something usually called 'vanity publishing'—are likely to be judged more harshly. We have therefore to conclude that if the RAE had existed in Germany in the 1880s Frege would have got a very low rating.

Even in the RAE free Germany of the period, Frege did not have an easy time. A fascinating portrait of him in the years 1910–14 is given by Carnap in his intellectual autobiography (Carnap, 1963). Carnap's involvement with Frege appears to have come about rather by chance. Carnap's family lived in Jena and Carnap went to the local university where Frege taught. Carnap writes (1963, 5):

> In the fall of 1910, I attended Frege's course "Begriffsschrift" (conceptual notation, ideography), out of curiosity, not knowing anything either of the man or the subject except for a friend's remark that somebody had found it interesting. We found a very small number of other students, there. Frege looked old beyond his years. He was of small stature, rather shy, extremely introverted. He seldom looked at the audience. Ordinarily we saw only his back, while he drew the strange diagrams of his symbolism on the blackboard and explained them. Never did a student ask a question or make a remark, whether during the lecture or afterwards. The possibility of a discussion seemed to be out of the question.

Earlier in his account, Carnap says (1963, 4):

> Gottlob Frege (1848–1925) was at that time, although past 60, only Professor Extraordinarius (Associate Professor) of mathematics in Jena. His work was practically unknown in Germany; neither mathematicians nor philosophers paid any attention to it.

It was obvious that Frege was deeply disappointed and sometimes bitter about this dead silence.

Carnap, however, took a liking to Frege's work and attended his two advanced courses "Begriffsschrift II" in 1913 and his course Logik in der Mathematik in 1914. Carnap records that "Begriffsschrift II" was attended by 3 students: Carnap, a friend of Carnap's, and (Carnap, 1963, 5) 'a retired major of the army who studied some of the new ideas in mathematics as a hobby.'

We can see from this that Frege's career was hardly a great success, but, if there had been a RAE regime in Germany, things would have gone even worse for him. As we have seen, Frege would undoubtedly have got a low rating in the RAE exercise, and the inevitable penalties would have fallen on his head. His research time would have been cut and he would have been forced to take on extra teaching duties. Thus he would not have had the necessary research time to develop his mathematical logic. Moreover, as we can see from Carnap's description, Frege may not have performed particularly well as a teacher. He seemed to attract very few students, and his teaching technique does not appear to have been of the kind recommended by educational experts. Having failed as both a researcher and a teacher, there is little doubt than, under a RAE regime, Frege would have been forced to retire early rather than allowed to stay on until he was 70. Thus Carnap would never have been able to attend his lectures, and the development and diffusion of the new important ideas of mathematical logic would have been held up still further.

4. Second Case-History: Semmelweis and Antisepsis

My second case-history, as we shall see, has many points in common with the first. However, it does differ very strikingly as regards the branch of science in which the research was conducted. Frege's research was purely theoretical, and was carried out in a branch of mathematics, mathematical logic, which is closely linked to philosophy. Semmelweis's research by contrast was highly empirical, and was carried out in medicine. In fact Semmelweis's investigation was into the causes of a terrible disease (puerperal fever) which affected women who had just given birth. Puerperal fever was, at the time, the principal cause of death in childbirth.

Semmelweis was Hungarian, but studied medicine at the University of Vienna. In 1844 he qualified as a doctor, and, later in the same year obtained the degree of Master of Midwifery. From then until 1849, he held the posts of either aspirant to assistant or full assistant at the first maternity clinic in Vienna. It was during this period that he carried out his research.[2]

The Vienna Maternity Hospital was divided into two clinics from 1833. Patients were admitted to the two clinics on alternate days thereby producing, unintentionally, a system of random allocation. Between 1833 and 1840, medical students, doctors and midwives attended both clinics, but, thereafter, although doctors went to both clinics, the first clinic only was used for the instruction of medical students who were all male in those days, and the second clinic was reserved for the instruction of midwives. When Semmelweis began working as a full assistant in 1846, the mortality statistics showed a strange phenomenon

Between 1833 and 1840, the death rates in the two clinics had been comparable, but, in the period 1841–46, the death rate in the first clinic was 9.92% and in the second clinic 3.88%. The first figure is more than 2.5 times the second—a difference which is certainly statistically significant. The quoted figures actually underestimate the difference since some severe cases of puerperal fever were removed from the first clinic to the general hospital where they died—thereby disappearing from the first clinic's mortality statistics. This rarely happened in the second clinic. Semmelweis was puzzled and set himself the task of finding the cause of the higher death rate in the first clinic.

Semmelweis followed a procedure rather similar to Popper's conjectures and refutations. He considered in turn a number of hypotheses as to what might be the cause of the difference between the two clinics. He then compared these hypotheses to the facts, and found that each one of a long series of hypotheses was refuted by this comparison. Eventually, however, Semmelweis did hit on a hypothesis which was corroborated by the observations.

The first hypothesis considered by Semmelweis was that the higher death rate in the first clinic was due to 'atmospheric-cosmic-terrestial' factors. This sounds strange but is just a way of referring to the miasma theory of disease which was standard at the time. However Semmelweis pointed out that it could not explain the

---

[2] This account of Semmelweis's research is a shortened version of the one given in my paper: Gillies (2005). That paper also contains more detailed references to the considerable literature on Semmelweis. Semmelweis's own account of his researches in Semmelweis (1861) is also worth consulting.

different mortality rates in the first and second clinics. These were under the same roof and had an ante-room in common. So they must be exposed to the same 'atmospheric-cosmic-terrestial' influences. Yet the death rates in the two clinics were very different.

The next hypothesis was that overcrowding was the key factor, but this too was easily refuted since the second clinic was always more crowded than the first, which, not surprisingly had acquired an evil reputation among the patients, almost all of whom tried to avoid it.

In this sort of way Semmelweis eliminated quite a number of curious hypotheses. One concerned the appearance of a priest to give the last sacrament to a dying woman. The arrangement of the rooms meant that the priest, arrayed in his robes, and with an attendant before him ringing a bell had to pass through five wards of the first clinic before reaching the sickroom where the woman lay dying. The priest had, however, direct access to the sickroom in the case of the second clinic. The hypothesis then was that the terrifying psychological effect of the priest's appearance debilitated patients in the first clinic, and made them more liable to puerperal fever. Semmelweis persuaded the priest to come by a less direct route, without bells, and without passing through the other clinic rooms. The two clinics were made identical in this respect as well, but the mortality rate was unaffected.

After trying out these hypotheses and others unsuccessfully, Semmelweis was in a depressed state in the winter of 1846–7. However a tragic event early in 1847 led him to formulate a new hypothesis. On 20th March 1847, Semmelweis heard with sorrow of the death of Professor Kolletschka. In the course of a post-mortem examination, Professor Kolletschka had received a wound on his finger from the knife of one of the students helping to carry out the autopsy. As a result Kolletschka died not long afterwards of a disease very similar to puerperal fever. Semmelweis reasoned that Kolletschka's death had been owing to cadaverous matter entering his bloodstream. Could the same cause explain the higher death rate of patients in the first clinic? In fact professors, assistants and students often went directly from dissecting corpses to examining patients in the first clinic. It is true that they washed their hands with soap and water, but perhaps some cadaverous particles still adhered to their hands. Indeed this seemed probable since their hands often retained a cadaverous odour after washing. The doctors and medical students might then infect some of the patients in the first clinic with these cadaverous particles, thereby giving them puerperal fever. This would explain why the death rate was lower in the second clinic, since the student midwives did not carry out post-mortems.

In order to test this hypothesis, Semmelweis, from some time in May 1847, required everyone to wash their hands in disinfectant before making examinations. At first he used *chlorina liquida*, but, as this was rather expensive, chlorinated lime was substituted. The result was dramatic. In 1848 the mortality rate in the first clinic fell to 1.27%, while that in the second clinic was 1.30%. This was the first time the mortality rate in the first clinic had been lower than that of the second clinic since the medical students had been divided from the student midwives in 1841.

Through a consideration of some further cases, Semmelweis extended his theory to the view that, not just cadaverous particles, but any decaying organic matter, could cause puerperal fever if it entered the bloodstream of a patient.

Let us now look at Semmelweis's theory from a modern point of view. Puerperal fever is now known as 'post-partum sepsis' and is considered to be a bacterial infection. The bacterium principally responsible is *streptococcus pyogenes*, but other *streptococci* and *staphylococci* may be involved. Thus, from a modern point of view, cadaverous particles and other decaying organic matter would not necessarily cause puerperal fever but only if they contain a large enough quantity of living *streptococci* and *staphylococci*. However as putrid matter derived from living organisms is a good source of such bacteria, Semmelweis was not far wrong.

As for the hand washing recommended by Semmelweis, that is of course absolutely standard in hospitals. Medical staff have to wash their hands in antiseptic soap (hibiscrub), and there is also a gelatinous substance (alcogel) which is squirted on to the hand. Naturally a doctor's hands must be sterilised in this way before examining any patient—exactly as Semmelweis recommended.

Not only are Semmelweis's views regarded as largely correct form a modern point of view, but the investigation which led him to them is now held up as model of good scientific method. In fact Hempel in his 1966 book: *Philosophy of Natural Science* gives a number of examples of what he regards as excellent scientific investigations, and the very first of these is Semmelweis's research into puerperal fever.

This then is the modern point of view, but how did Semmelweis's contemporaries react to his new theory of the cause of puerperal fever and the practical recommendations based on it?

Lessons from the History and Philosophy of Science

The short answer is that Semmelweis's reception by his contemporaries was almost exactly the same as Frege's. Semmelweis did manage to persuade one or two doctors of the truth of his findings, but the vast majority of the medical profession rejected his theory and ignored the practical recommendations based upon it. I discuss some of the detailed responses to Semmelweis in my longer paper on the subject (see Gillies, 2005, 178–9). Here I will only mention one typical reaction. After Semmelweis had made his discovery in 1848, he and some of his friends in Vienna wrote about them to the directors of several maternity hospitals. Simpson of Edinburgh replied somewhat rudely to this letter saying that its authors obviously had not studied the obstetrical literature in English. Simpson was of course a very important figure in the medical world of the time. He had introduced the use of chloroform for operations, and had recommended its use as a pain-killer in childbirth. His response to Semmelweis and his friends is very similar in character to Venn's review of Frege's *Begriffsschrift*.

In Vienna the Professor and Head of the Maternity Clinics, Johann Klein, was opposed to Semmelweis's ideas, and his opposition, and that of others, caused Semmelweis to leave Vienna in 1850. He did however get a position in a Maternity Hospital at Budapest in his native Hungary. Here he wrote up his new theory of the causes of puerperal fever, and answered the objections which had been made to it. These writings were published in book form in 1861, but once again had no success in persuading the medical profession to adopt his ideas.

Semmelweis's case is very similar to Frege's. Semmelweis, like Frege, had great difficulties, and, if counterfactually, there had been a RAE regime at the time, these difficulties would have become worse. Semmelweis's work would obviously have been judged by peer review to have no value, and his allowance of research time would have been reduced, so that he might not have had the time to write up his results in book form and to answer his critics.

The failure of the research community to recognise Semmelweis's work had of course much more serious consequences than the corresponding failure to appreciate Frege's innovations. In the twenty years after 1847 when Semmelweis made his basic discoveries, hospitals throughout the world were plagued with what were known as 'hospital diseases', that is to say, diseases which a patient entering a hospital was very likely to contract. These included not just puerperal fever, but a whole range of other unpleasant illnesses. There were wound sepsis, hospital gangrene,

Donald Gillies

tetanus, and spreading gangrene, erysipelas (or 'St. Anthony's fire'), pyaemia and septicaemia which are two different forms of blood poisoning, and so on. Many of these diseases were fatal. From the modern point of view, they are all bacterial diseases which can be conquered by applying the kind of antiseptic precautions recommended by Semmelweis.

In 1871, over twenty years after his rather abrupt reply to Semmelweis and his friends, Simpson of Edinburgh wrote a series of articles on 'Hospitalism'. These contained his famous claim, well-supported by statistics, that 'the man laid on the operating-table in one of our surgical hospitals is exposed to more chances of death than the English soldier on the field of Waterloo'. Simpson thought that hospitals infected with pyaemia might have to be demolished completely. So serious was the crisis, that he even recommended replacing hospitals by villages of small iron huts to accommodate one or two patients, which were to be pulled down and re-erected periodically. Luckily the theory and practice of antisepsis were introduced in Britain by Lister in 1865, and were supported by the germ theory of disease developed by Pasteur in France and Koch in Germany. The new antiseptic methods had become general by the mid 1880s, so that the hospital crisis was averted. All the same, the failure to recognise Semmelweis's work must have cost the lives of many patients.

In my longer paper on the Semmelweis case (Gillies, 2005, 180–1), I argue that, in the history and philosophy of science, it is customary to cite historical examples of excellent science in order to exemplify what are claimed to be good methodological principles for science. However instances in which the scientific community makes a mistake, as happened in the Semmelweis case or that of Frege, can also be valuable in suggesting new rules of practice designed to make such mistakes less likely in the future. From this point of view, the RAE is clearly a step backwards. Instead of learning from the mistakes which were made regarding Frege and Semmelweis, and introducing a system designed to make such mistakes less likely in the future, it does the opposite. If the RAE had been in existence in the days of Frege and Semmelweis, it would, as we have seen, have made their position even worse than it already was. Naturally the same will apply to any future brilliant innovators like Frege and Semmelweis who have the misfortune to be working in a RAE regime.

This point can also be made by introducing a distinction taken from the theory of statistical tests. Statistical tests are said to be liable to two types of error (Type I error, and Type II error). A

Lessons from the History and Philosophy of Science

Type I error occurs if the test leads to the rejection of a hypothesis which is in fact true. A Type II error occurs if the test leads to the confirmation of a hypothesis which is in fact false. Analogously we could say that a research assessment procedure commits a Type I error if it leads to funding being withdrawn from a researcher or research programme which would have obtained excellent results had it been continued. A research assessment procedure commits a Type II error if it leads to funding being continued for a researcher or research programme which obtains no good results however long it goes on. This distinction leads to the following general criticism of the RAE. The RAE concentrates exclusively on eliminating Type II errors. The idea behind the RAE is to make research more cost effective by withdrawing funds from bad researchers and giving them to good researchers. No thought is devoted to the possibility of making a Type I error, the error that is of withdrawing funding from researchers who would have made important advances if their research had been supported. Yet the history of science shows that Type I errors are much more serious than Type II errors. The case of Semmelweis is a very striking example. The fact that his line of research was not recognised and supported by the medical community meant that, for twenty years after his investigation, thousands of patients lost their lives and there was a general crisis in the whole hospital system.

In comparison with Type I errors, Type II errors are much less serious. The worst that can happen is that some government money is spent with nothing to show for it. Moreover Type II errors are inevitable from the very nature of research. Suppose research is required on some problem, and there are four different approaches to its solution which lead to four different research programmes. It may be almost impossible to say at the beginning which of the four programmes is going to lead to success. Suppose it turns out to be research programme number 3. The researchers on programmes 1, 2 & 4 may be just as competent and hard-working as those on programme 3, but, because their efforts are being made in the wrong direction, they will lead nowhere. Suppose programme 3 is cancelled in order to save money (Type I error), then all the money spent on research in the problem will lead nowhere. It will be a total loss. On the other hand if another programme (5) is also funded, the costs will be a bit higher but a successful result will be obtained. This shows why Type I errors are much more serious than Type II errors, and why funding bodies should make sure that some funding at least is given to every research school and approach rather than concentrating on the hopeless task of trying to foresee which approach will in the long run prove successful.

The same analysis also shows why peer review as a system can often give wrong results. Let us return to our example of the problem being tackled by four different research programmes, of which programme number 3 ultimately proves successful. Let us suppose further (which indeed is often the case) that initially programme 3 attracts many fewer researchers than programmes 1, 2 & 4. Now it is characteristic of most researchers that they think their own approach to the problem is the correct one, and that other approaches are misguided. If a peer review is conducted by a committee whose researchers are a random sample of those working on the problem, then the majority will be working on programmes 1, 2 & 4, and are therefore very likely to give a negative judgement of the merits of programme 3. As the result of the recommendation of such a peer review, funding might be withdrawn from programme 3, and the solution of the problem might remain undiscovered for a long time.

5. Third Case-History: Copernicus and Astronomy

I now turn to my third example which I will deal with more concisely both because it is more familiar and because my general line of argument should by now have become fairly clear. However, it is worth looking at this example because it deals with yet another branch of science (astronomy) and also a different historical period.[3]

Copernicus (1473–1543) was born in which is now Poland and studied at Universities in both Poland and Italy. Through the influence of his uncle, he obtained the post of Canon of Frauenberg Cathedral in 1497, and held this position until his death. Copernicus' duties as canon seem to have left him plenty of time for other activities, and he seems to have devoted much of this time to developing in detail his new theory of the universe. This was published as *De Revolutionibus Orbium Caelestium*, when Copernicus was on his death bed. In the preface Copernicus states that he had meditated on this work for more than 36 years.

There is little doubt that during Copernicus' lifetime and for more than 50 or 60 years after his death, his view that the Earth

[3] A more detailed account of Copernicus' work on astronomy is to be found in Kuhn (1957).

moved was regarded as absurd, not only by the vast majority of the general public, but also by the vast majority of those who were expert in astronomy. It is significant that *De Revolutionibus* was not put on the index by the Roman Catholic Church until 1616. It was not until then that Copernicanism had sufficient adherents to be considered a threat.

Although the majority of expert astronomers of the period would have dismissed the Copernican view as absurd, a few such astronomers, notably Kepler and Galileo, did side with Copernicus and carried out researches developing his theory until, in due course, it won general acceptance by astronomers not influenced by the Roman Catholic Church's opposition.

Copernicus' research, like that of Frege and Semmelweis, had very important practical applications. Despite the Roman Catholic Church's opposition to his theory, his calculations were used in the reform of the calendar carried out by Pope Gregory XIII in 1582. Ironically the Protestant countries, whose astronomers were the first to accept the Copernican theory, rejected the Gregorian calendar for a long time on the ground that it had been introduced by the Roman Catholic Church and must presumably therefore be bad. Copernicus' theory was also used to produce improved astronomical tables. Reinhold used *De Revolutionibus* in the production of his Prutenic Tables which appeared in 1551. These were the first complete tables prepared in Europe for three centuries. In 1627, they were superseded by the Rudolfine Tables which Kepler produced using his much improved version of Copernicus's theory. The Rudolfine Tables were clearly superior to all astronomical tables in use before. Of course astronomical tables were applied in navigation, and so were an important tool for promoting the growth of European maritime trade.

Let us now once again apply our counterfactual methodology and consider how Copernicus would have been affected if, instead of being a Canon of Frauenberg, he had lived under a RAE regime. Of course it is indeed rather anachronistic to suppose that something like the RAE might have been in existence in such a distant historical period. Yet I think we can still say with some confidence that if it had existed then, it would have impacted negatively on Copernicus. Under a RAE regime, Copernicus would not have been allowed to continue his research peacefully as a Canon of Frauenberg. On the contrary, he would have been brought to account to make sure he was not wasting the tax-payers' money. In order to be allowed time to continue his research, he would have had to submit samples of his research work to a committee of experts in the field. Now nearly all these experts, as we have already pointed out, would have judged that Copernicus' research was absurd and not worth funding. Thus Copernicus would have been sent off to a teaching university with little time for research, and would have had to devote most of his time to teaching astronomy to undergraduates. Naturally, as the syllabus would have been determined by the majority of his colleagues, he would have had to teach, not his new theory, but the standard Aristotelian-Ptolemaic account of astronomy. So Copernicus, under a system of funding of RAE type, would have been deprived of his research time, and forced to spend his days teaching the Aristotelian-Ptolemaic account of astronomy. Meanwhile the leading experts of the Aristotelian-Ptolemaic theory would have had posts at the well-funded research universities giving them plenty of time to pursue their research. No doubt they would have developed Aristotelian-Ptolemaic theory by means of ever more mathematically ingenious combinations of epicycles. It need hardly be said that all this would have acted as an extreme dampener on the progress of astronomy.

I have given three examples of cases in which a regime of RAE type would have impeded rather than helped scientific advance. Of course many more cases along the same lines could be described, but it will now be more fruitful to turn from history of science to philosophy of science. In the next section, I will try to analyse the factors, which in the cases of Copernicus, Frege and Semmelweis, led to the failure of the peer review method. As we shall see, this is not a problem which has arisen in just a few cases, but is an underlying pattern in the development of science.

6. Kuhn's Distinction between Normal and Revolutionary Science

The part of philosophy of science which I would like to consider is Kuhn's theory of scientific development as set out in his *The Structure of Scientific Revolutions* (1962). Kuhn's view is that science develops through periods of *normal science* which are characterised by the dominance of a *paradigm*, but which are interrupted by occasional revolutions during which the old paradigm is replaced by a new one. Kuhn gives three main examples of scientific revolutions. These are the Copernican

Revolution, the Chemical Revolution, and the Einsteinian Revolution. As we have already discussed Copernicus, I will illustrate Kuhn's views by a brief account of the other two examples.

The Chemical Revolution. The main theme of the chemical revolution was the replacement of the *phlogiston* theory by the *oxygen* theory, though there were many other important changes as well. According to the phlogiston theory, bodies are inflammable if they contain a substance called phlogiston, and this is released when the body burns. The phlogiston theory was also used to explain the calcination of metals. When a metal is heated in air, in many cases it turns into a powder known as the *calx*, e.g. iron → rust. Conversely the calx is usually found in ores of the metal, and the metal itself could often be obtained by heating with charcoal. These transformations were explained by postulating that

calx + phlogiston = metal

When we heat a metal, phlogiston is given off, and the calx remains. Conversely when we heat the calx with charcoal, since charcoal is very rich in phlogiston because it burns easily, the phlogiston from the charcoal combines with the calx to give the metal.

In the oxygen theory, burning is explained as the combination of the substance with oxygen; while the calx is identified with the oxide of the metal. So turning a metal into its calx by heating in air is explained by the equation

metal + oxygen = metal oxide

Similarly obtaining the metal by heating the calx with charcoal is explained by the equation

metal oxide + carbon = metal + carbon dioxide

The oxygen theory was developed by Lavoisier. At the beginning of his researches in 1772, he was already sceptical of the then dominant phlogiston theory. In the next decade or so, many experimental discoveries concerning gases were made. These discoveries were mainly owing to the English experimental chemists—particularly Priestley and Cavendish. However, these English chemists remained faithful to the phlogiston theory. For example Priestley referred to what we now call oxygen as dephlogisticated air. Lavoisier, on the other hand, reinterpreted their results in terms of his new and developing oxygen theory. Lavoisier's new paradigm for chemistry was set out in his *Traité élémentaire de chimie* of 1789, and within a few years it was adopted

Donald Gillies

by the majority of chemists. Priestley, however, who lived until 1804 never gave up the phlogiston theory.

The Einsteinian Revolution. The triumph of the Newtonian paradigm initiated a new period of normal science for astronomy (c. 1700—c. 1900). The dominant paradigm consisted of Newtonian mechanics including the law of gravity, and the normal scientist had to use this tool to explain the motions of the heavenly bodies in detail—comets, perturbations of the planets and the moon, etc. In the Einsteinian revolution (c. 1900—c. 1920), however, the Newtonian paradigm was replaced by the special and general theories of relativity.

Further research in the philosophy of science has shown that Kuhn's model, with some modifications, can be extended to mathematics and medicine. Thus Frege's work can be considered as a initiating a revolution in logic analogous to the Copernican revolution in astronomy. The change was from an Aristotelian paradigm, whose core was the theory of the syllogism, to a new paradigm whose core was propositional and first-order predicate calculus.[4] Then again Semmelweis's investigation can be seen as one of the first steps in a revolution in medicine. The change was from a paradigm whose core was the miasma and contagion theories of disease to a new paradigm with the germ theory of disease as its core.[5]

Now one of the strengths of Kuhn's theory is that it explains why the scientific community made such mistaken judgements regarding figures like Copernicus, Semmelweis and Frege. On Kuhn's model, at the beginning of a revolution almost all the researchers in the field accept the dominant paradigm, and, from the point of view of this paradigm, the new revolutionary approach will indeed seem absurd.

Another important consequence of Kuhn's theory is that the mistaken judgements regarding Copernicus, Semmelweis and Frege are not features of science's past, but are likely to recur over and over again. Of course, long before Kuhn, the Copernican revolution had been studied by historians of science. However, it tended to be regarded as something of a 'one-off' event—a dramatic change which had introduced modern science, but was not likely to recur. This is reflected in the fact that it was often referred to as: *The* Scientific Revolution. Kuhn's originality was to suggest that all branches of science develop through periodic revolutions.

---

[4] For more details, see Gillies (1992).
[5] For more details, see Gillies (2005).

This new view was obviously suggested by the revolution in physics in the first few decades of the twentieth century which led to the triumph of relativity theory and quantum mechanics. Kuhn's model of scientific development was roughly as follows. For most of the time we have 'normal science' in which the scientists working in a particular area all, except perhaps for a few dissidents, accept the same dominant paradigm. Within the framework of that paradigm, steady, if perhaps somewhat slow, progress is made. Every so often, however, a period of revolution occurs in which the previously dominant paradigm comes to be criticized by a small number of revolutionary scientists. This small group succeeds in developing a new paradigm, and in persuading their colleagues to accept it. Thus there comes about a revolutionary shift from the old paradigm to a new one. Although revolutions occur only occasionally in the development of a field of science, such revolutions are the exciting times in which really big progress is made in the field.

Kuhn's model of scientific development is, in my view, strongly confirmed by studies in the history of science. Indeed it applies not just to the natural sciences considered by Kuhn, but also to science in the broader sense considered in this paper which includes also mathematics and medicine. In the next section, I will use Kuhn's model to examine the likely effects of the RAE on scientific research in the UK.

7. Analysis of the Likely Effects of the RAE

Let us begin by considering the effects of the RAE on normal science. In a period of normal science, those working in a branch of the subject will all accept the dominant paradigm, and no revolutionary alternative will have been suggested. It will then be an easier matter for the experts in the field to judge who is best according to the criteria of the dominant paradigm. Allocating research funding to these most successful 'puzzle solvers', as Kuhn calls them, will usually enable the normal science activity of puzzle solving to continue successfully. One qualification to this must, however, be introduced on the basis of the discussion of Type I and Type II errors which was given at the end of section 4. We there gave an example of research into a problem, where there are four different approaches to its solution leading to four different research programmes. This situation is still possible, and indeed often occurs, in normal science, for the four different research programmes could all be compatible with the dominant paradigm. As we pointed out, in such circumstances, a thoughtless use of peer review as a tool could easily lead to wrong decisions. Suppose programme 3 in fact turns out to be the one which leads to the solution of the problem, but suppose initially it is supported by only a few researchers. A peer review conducted by a committee chosen at random from those working on the problem might well contain an overwhelming majority of researchers working on programmes 1, 2 & 4, and such a committee could easily recommend the cancellation of funding for research programme 3, a decision which would have disastrous long term results. With this qualification, however, we can say that the RAE is not likely to have too damaging an effect on normal science. The only problem is that normal science tends to be routine in character and to produce small advances rather slowly. Surely, however, we want a research regime to encourage big advances in the subject, exciting innovations, breakthroughs, etc.

It is precisely here that the RAE is likely to fail. Any big advance is likely to have something revolutionary about it, something which challenges accepted ideas and paradigms. However it is precisely in these case, as we have shown above, that the RAE with its excessive reliance on peer review is likely to have a very negative effect. Our conclusion then is the RAE is likely to shift the UK research community in the direction of producing the routine research of normal science resulting in slow progress and small advances. At the same time it will have the effect of tending to stifle the really good research—the big advances, the exciting innovations, the major breakthroughs. Clearly then the overall effect of the RAE is likely to be very negative as regards research output in the UK.

The RAE is also likely to impact very negatively on the production of wealth-generating science-based technologies in the UK. The reason for this is that the most striking technologies from the point of view of wealth-generation are often based on revolutionary scientific advances. This is well-illustrated by the three examples considered in this paper. Copernicus' new astronomy led, as we have seen, to a much improved navigation, and this was essential to the profitable development of European sea-borne trade in the 17th and 18th centuries. The new mathematical logic introduced by Frege was essential for the development of the computer. It is significant here that Bertrand Russell was one of the first to recognise and develop Frege's work. Russell established an interest in mathematical logic in the UK, which passed on to two later researchers at Cambridge: Max

Newman and his student Alan Turing. After the Second World War, Newman and Turing were part of the team at Manchester which produced the Manchester Automatic Digital Machine (MADM). This started running in 1948, and can be considered as the first computer in the modern sense.[6] Thus Russell's early recognition of Frege's revolutionary innovations led indirectly to the UK taking an early lead in the computer field. This early lead was later lost, as we know, but this was owing to lack of sufficient investment by either the public or private sectors. There was no problem with the UK's research community in those pre-RAE days. Our third case was concerned the revolutionary introduction of antisepsis in conjunction with revolutionary new theories about the causes of disease. We focussed on Semmelweis whose research work was rejected by the medical community of his time. As we remarked, however, Lister was more successful, and was able to persuade the medical community in the UK to accept antisepsis. This was obviously of great benefit to patients, but I would now like to add that it led to very successful business developments. For his new form of surgery Lister needed antiseptic dressings, and he devoted a lot of time and thought to working out the best design and composition of such dressings. As his ideas came to be accepted, the demand for these dressings increased and companies were formed to produce them. One of these was founded by a pharmacist Thomas James Smith. In 1896, he went into partnership with his nephew Horatio Nelson Smith to produce and sell antiseptic dressings. They called the firm Smith and Nephew. Today Smith and Nephew is a transnational company operating in 33 countries and generating sales of £1.25 billion. The company is still involved in wound care as one of its three main specialities, but it has expanded into orthopaedics and endoscopy. One of its well-known products is elastoplast which was developed in 1928. The general design of elastoplast is based on some of the antiseptic dressings developed by Lister. The commercial success of Smith and Nephew is a good illustration of the importance of having a satisfactory research regime in the UK. If Lister's research on antisepsis had met the same fate as that of Semmelweis only 17 years earlier, then the firm of Smith and Nephew would not be with us today.

---

[6] For more details about the Manchester Automatic Digital Machine and its claim to be the first computer in the modern sense, see Gillies and Zheng, 2001, 445–9.

Donald Gillies

8. General Conclusions

The RAE is very expensive both in money and in the time which academics in the UK have to devote to it. I have argued in this paper that its likely effect is to shift the UK research community in the direction of producing the routine research of normal science resulting in slow progress and small advances, while tending to stifle the really good research—the big advances, the exciting innovations, the big breakthroughs. Thus a great deal of tax payers' money is being spent on an exercise whose likely effect is to make the research output of the UK worse rather than better. Only one conclusion can be drawn from this, namely that the RAE should be abolished straightaway.

My general argument has brought to light three major faults in the RAE. (1) The RAE rules out the research strategy of working for many years on a piece of research before publication. Yet this strategy has proved very successful in the past. We gave Copernicus and Wittgenstein as examples of the success of this strategy, but many other examples could of course be given. (2) The RAE relies too strongly on peer review, which may work not too badly for normal science, but which can give very erroneous results when it comes to the most important revolutionary advances in science. Frege, Semmelweis and Copernicus were all examples of this. (3) The RAE concentrates too much on trying to eliminate Type II error, that is the error of funding bad research, but devotes no consideration to eliminating Type I error, that is the error of failing to fund good research. Yet Type I errors have much more damaging effects on the progress of research than Type II errors. This was illustrated above all by the case of Semmelweis where a Type I error of failing to recognise and support important research led to thousands of patients dying and a general crisis in the hospitals.

Abolition of the RAE could be accomplished very easily because it would produce no disruption in the system. Indeed research in the UK was very successful for many decades with no RAE. We argued for this in detail in the case of philosophy but the same applies to other areas of research. However, here it might be objected that we can't just go back to the *status quo ante* RAE, for the whole university system and research community has expanded considerably since 1975, and so can no longer be run along the lines which were used in this earlier period. I agree with this point, but would still stress that there is no hurry to introduce a new system for organising research, and that there should be a great deal of thought, discussion and consultation before doing so. What our critique of the RAE has shown is that research is rather a subtle and complicated activity and that producing a regime in which it flourishes is not an easy matter. Perhaps the biggest difficulty lies in the fact that we cannot tell immediately whether a piece of research is good, and sometimes it is only after a period of as long as thirty years that a fairly definite judgement can be reached. In this respect research differs very strikingly from competitive sports such as tennis or football. We can grade tennis players at a particular moment simply by getting them to play each other in tournaments and seeing who beats who. However, we cannot be sure that a researcher whose work is now judged to be of poor quality may not turn out after all to be a Copernicus, a Semmelweis, or a Frege. Even in cases where it is recognised that a scientific discovery has been made, the importance of that discovery may not become apparent for many years. A good example of this is Alexander Fleming's discovery of penicillin which was made in 1928, and published by Fleming in 1929. Fleming was not harshly treated like Semmelweis or Frege, but the significance of his discovery was certainly not recognised immediately. The head of the laboratory where Fleming worked (Sir Almroth Wright) was a Fellow of the Royal Society and a great admirer of Fleming. In 1930 Wright proposed Fleming for the Royal Society citing his discovery of penicillin and some other research achievements. Fleming, however, was not elected in 1930 or in the four following years. In fact Fleming only became a Fellow of the Royal Society in 1943 when he was 62 years old.[7]

But if it is not possible to tell whether a piece of research is good except after a long lapse of time—perhaps as long as thirty years, how can we possibly decide what research to fund at a given moment? It almost looks as if the problem of devising a sensible system for funding research is insoluble. This is not really so, however, and there are ways of overcoming the difficulty. Needless to say, however, they are not along the lines of the RAE. What is needed here is some new thinking and a quite different approach. I have my own ideas of what this different approach might be like, but it would not be appropriate to give them here since my aim in this paper is to give a critique of the existing system. I hope, however, that my paper might be useful to those trying to devise a new system for organising research in the UK by suggesting a way

[7] These details about Fleming and the Royal Society are taken from Macfarlane, 1984, 140–1, & 202.

Donald Gillies

in which their ideas can be tested. I would suggest that anyone who has thought out a possible research regime should consider a number of major research achievements of the past such as those of Copernicus, Frege, Semmelweis and Wittgenstein, and examine the effect that the proposed research regime would have had on those achievements. If the effect turns out to be negative, then the proposed research regime should be rejected and something better devised to replace it.

References

Anderson, A.R. (ed.) (1964) *Minds and Machines*, Prentice-Hall.

Bell, J.L. and Machover, M. (1977) *A Course in Mathematical Logic*, North-Holland.

Bynum, T.W. (ed.) (1972) *Gottlob Frege. Conceptual Notation and Related Articles,* Oxford University Press.

Carnap, R. (1963) Intellectual Autobiography. In P.A.Schilpp (ed.), *The Philosophy of Rudolf Carnap*, Library of Living Philosophers, Open Court, 3–84.

Davis, M. (1988a). Mathematical Logic and the Origin of Modern Computing. In Rolf Herken (ed.), *The Universal Turing Machine. A Half-Century Survey*, Oxford University Press, 149–74.

Davis, M. (1988b). Influences of Mathematical Logic on Computer Science. In Rolf Herken (ed.), *The Universal Turing Machine. A Half-Century Survey*, Oxford University Press, 315–26.

Edmonds, D. and Eidinow, J. (2001) *Wittgenstein's Poker. The Story of a Ten-Minute Argument between two Great Philosophers.* faber and faber.

Frege, G. (1879) *Begriffsschrift, eine der Arithmetischen nachgebildete Formelsprache des reinen Denkens.* English translation in Bynum (1972), 101–203.

Gillies, D.A. (1992) The Fregean Revolution in Logic. In D.A.Gillies (ed.) *Revolutions in Mathematics*, Oxford University Press, 265–305.

Gillies, D.A. (2002) Logicism and the Development of Computer Science. In Antonis C.Kakas and Fariba Sadri (eds.) *Computational Logic: Logic Programming and Beyond*, Part II, Springer, 588–604.

Gillies, D.A. (2005) Hempelian and Kuhnian approaches in the Philosophy of Medicine: the Semmelweis case, *Studies in History and Philosophy of Biological and Biomedical Sciences*, **36**, 159–181.

Gillies, D.A. and Zheng, Y. (2001) Dynamic Interactions with the Philosophy of Mathematics, *Theoria*, **16**, 437–59.

Hempel, C.G. (1966) *Philosophy of Natural Science*, Prentice-Hall.

Kuhn, T.S. (1957) *The Copernican Revolution*, Vintage, 1959.

Kuhn. T.S. (1962) *The Structure of Scientific Revolutions*, University of Chicago Press, 1969.

Lessons from the History and Philosophy of Science

Lucas, J.R. (1961) Minds, Machines and Gödel, *Philosophy*, **36**. Reprinted in Anderson, 1964, 43–59.

Macfarlane, G. (1984) *Alexander Fleming. The Man and the Myth*, Chatto & Windus

Malcolm, N. (1958) *Ludwig Wittgenstein. A Memoir*, 2nd Edition, Oxford University Press, 1989.

Mendelson, E. (1964) *Introduction to Mathematical Logic*, Van Nostrand.

Monk, R. (1990) *Ludwig Wittgenstein. The Duty of Genius*, Jonathan Cape.

Pitcher, G. (1964) *The Philosophy of Wittgenstein*, Prentice-Hall.

Semmelweis, I. (1861) *The Etiology, Concept, and Prophylaxis of Childbed Fever*. English Translation by K. Codell Carter, The University of Wisconsin Press, 1983.

Turing, A.M. (1950) Computing Machinery and Intelligence, *Mind*, **59**. Reprinted in Anderson, 1964, 4–30.

Wittgenstein, L. (1953) *Philosophical Investigations*. English Translation by G.E.M.Anscombe, Blackwell, 1963.

The Ravens Revisited

PETER LIPTON

> And here it is constantly supposed, that there is a connection between the present fact and that which is inferred from it. Were there nothing to bind them together, the inference would be entirely precarious.
>
> *David Hume[1]*

Astronomers study the behaviour of the stars; philosophers of science study the behaviour of the astronomers. Philosophers of science, alongside historians and sociologists of science, are in the business of accounting for how science works and what it achieves. There is more to the philosophy of science than principled descriptions of scientific activity, since there are also all the normative questions of justification and warrant, but the descriptive task is an important part of the discipline and the primary focus of the present essay.

Science is one of the most complex and impressive things our species does, so it is natural to wish to understand how it works. Worthwhile but relatively easy, one might think. For unlike stars, astronomers can talk. To find out how science works, it might seem that all that is required is to get a few cooperative scientists to describe what they do. But things are not so easy, in part because there is such a large gap between what people can do and what they can describe. It is one thing to be able to ride a bicycle; it something quite different to be good at describing the physics and the physiology behind that ability. The same contrast between doing and describing applies to more sedentary pursuits. It is one thing to be expert at distinguishing grammatical from ungrammatical strings of words in one's native tongue; it is something quite different to be able to specify the principles by which these discriminations are made. The same applies to science. It is one thing to be a good scientist; it is something quite different to be good at giving a general description of what scientists do.

---

[1] D. Hume, *An Enquiry Concerning Human Understanding*, T. Beauchamp (ed.) (Oxford: Oxford University Press, 1748/1999), section IV.

Peter Lipton

Scientists are not good at the descriptive task. This is no criticism, since their job is to do the science, not to talk about how they do it. Philosophers of science are not very good at describing science either: this is more embarrassing, since it is their job. It turns out that giving even an approximately correct general description of what are apparently the simplest aspects of scientific research and judgement is remarkably difficult. For example, philosophers of science have worked long and hard to provide a general account of the apparently simple tripartite distinction between data that would be taken to support a given hypothesis, data that would tell against it, and data that would be irrelevant; but at least most of these attempts turn out to have the highly implausible consequence that the set of data irrelevant to a given hypothesis, the set one would have thought was the largest of the three by some margin, is empty. This difficulty is brought out with particular force by the raven paradox of confirmation, a problem that will occupy us for most of this essay.

Problems of Induction

The great Humean problem of induction is in the first instance about justification rather than description.[2] Our beliefs about things in the world we have not observed are based on inconclusive reasons. Hume's description of our inferential practice is glib: according to him, we suppose that the unobserved is like the observed. His main question is not what our practice is like, but what entitlement we have for following it. Given his description of that practice, for him the question of entitlement is tantamount to the question of what reason we have to believe that the unobserved is similar to the observed, what reason we can have to believe that nature is uniform. Hume's notorious answer is that no reason for the uniformity of nature is possible, because the uniformity claim is itself about the unobserved. In particular, observation could not teach us that the unobserved is like the observed, since observation could only teach us about the unobserved if we already knew that the unobserved is like the observed. You can't get there unless you have already arrived.

[2]   D. Hume, *An Enquiry Concerning Human Understanding*, T. Beauchamp (ed.) (Oxford: Oxford University Press, 1748/1999), sections IV, V.

The problem of induction is a problem of epistemic disconnection. If Hume is right, what we see gives us no guidance about what we have not seen. The observation of a black raven gives no more warrant for saying that the next raven will be black than for saying it will be blue. Every conjunction is arbitrary. Of course that is not the way we think: we use the observed as a guide to the unobserved, and we distinguish arbitrary conjunctions from those where the first conjunct is a guide to the second. Of course the probability of an arbitrary conjunction will go up if we observe what was before an uncertain conjunct, but that is mere 'content cutting', where discovering the truth of one conjunct does not affect the probability of the other. Genuine confirmation, by contrast, occurs when observing one conjunct increases the probability of the other.[3] Hume's sceptical argument is that content cutting is all we ever get, and that we have no basis for giving any determinate probability to the unobserved. But that is not at all the way we think. As Hume emphasised, we think that observing a black raven increases the probability that the next raven will be black: we think there is such a thing as inductive confirmation. We distinguish what we regard as arbitrary conjunctions—where one conjunct tells us nothing about another—from somehow connected conjuncts, where one gives information about the other. Our inductive practices make a clear distinction between genuine confirmation and mere 'content cutting'. We start by knowing part of the content of a statement is correct; in some cases we take that to provide a reason to believe that the rest of the content is correct too; in other cases we do not. Whether or not this distinction is justifiable, whether or not there can be any justification for using the observed as a guide to the unobserved, the descriptive question remains. On what basis do we make this discrimination between genuine confirmation and mere content cutting?

It is perhaps surprising that, like the problem of justification, the project of describing our inductive practices also throws up a problem of disconnection. Here the problem is in a way the mirror image of the problem of justification, for it is the problem of coming up with an account of confirmation that is not wildly over-permissive, incorrectly counting mere content cutting as if it were genuine confirmation. On initially plausible accounts, confirmation would be much too easy, so that A would count as confirming A&B however unrelated the two conjuncts may be. The

[3] N. Goodman, *Fact, Fiction and Forecast*, 4th ed. (Indianapolis: Bobbs-Merrill, 1983), 68–69.

challenge is to find an account that reflects the distinction we make in practice and so to avoid counting mere content cutting as confirmation, and to make sense of real confirmation.

The problem of arbitrary conjunctions arises immediately for a simple version of the hypothetico-deductive model of confirmation. On that model, the truth of the observed consequences of a hypothesis confirms the unobserved consequences. As a description this seems corroborated by a great deal of scientific practice. But it is an immediate consequence of the model that every statement counts as confirming any other, since for any conjunction A&B, A will count as confirming B.

An obvious response to this difficulty is to place a restriction on the logical form of the hypothesis to be confirmed, a restriction that rules out arbitrary conjunctions. This is what Nicod's criterion does, insisting that the hypothesis to be confirmed take the form not of a conjunction but of a universal conditional, such as 'All ravens are black' or 'All emeralds are green'.[4] But then along came Nelson Goodman, whose new riddle of induction supplied a nefarious mechanism for converting arbitrary conjunctions into the syntactic form of universal conditionals.[5] Goodman's famous example is 'All emeralds are grue', where the novel predicate refers to green things observed before a specified future time and blue things not so observed. Varying his example slightly, we may start with the arbitrary conjunction 'observed emeralds are green and unobserved ravens are black'. Green emeralds obviously do not confirm the prediction that unobserved ravens will be black, and Nicod's criterion does not immediately fall into the trap of saying they do, since this is a conjunction, not a universal conditional. But what Goodman taught us is that, by dint of modest creative predicate construction, this conjunction can be recast into the universal condition 'All emeravens are grack', and Nicod's criterion succumbs after all, mistakenly counting green emeralds as confirmation of black ravens.

The Raven Paradox

The other notorious puzzle of confirmation, the raven paradox, poses a similar challenge for the description of our inductive

[4] C. Hempel, *Aspects of Scientific Explanation* (New York: Free Press, 1965), 10–13.

[5] N. Goodman, *Fact, Fiction and Forecast*, 4th ed. (Indianapolis: Bobbs-Merrill, 1983), chapter III.

practices.[6] The problem is generated by the irresistible pull of the equivalence condition on confirmation, according to which whatever observation confirms an hypothesis must also confirm any logically equivalent hypothesis. (Irresistible, because if we see that two statements are logically equivalent, we see that they must stand or fall together, so any evidence for one is immediately evidence for the other.) Nicod's criterion says that 'All ravens are black' is confirmed by black ravens, and that 'All non-black things are non-ravens' is confirmed by things that are neither ravens nor black. But since those two hypotheses are logically equivalent—positive and contrapositive—the observation of green emeralds and white pieces of paper now count as if they too confirmed the hypothesis that all ravens are black. Once again, something has gone badly wrong. Taken together, the new riddle induction and the raven paradox reveal a lack of discrimination in our accounts of confirmation from both sides. The new riddle shows the evidence confirming too many hypotheses; the raven paradox shows a hypothesis being confirmed by too much evidence.

A common first reaction to the raven paradox is that we have somehow failed to register the obvious fact that the hypothesis that all ravens are black is only about ravens, and so only the colouration of ravens bears on its truth value. But as logical sophisticates know, a universal conditional actually makes a claim about every object in the universe. Another logical equivalent to 'All ravens are black' is 'Everything is either black or not a raven'. If anything in the universe fails to meet this specification, the raven hypothesis is false. Thus the paradox seems to show not just that some conjunctions are not confirmed by any of their conjuncts, but that confirmable conjunctions are in a sense not confirmed by all of their conjuncts. The raven hypothesis is something like a conjunction across every object in the universe: it claims that each of them has the property of being either black or not a raven. But what is the difference between those conjuncts that confirm and those that don't? How does the green emerald differ from the black raven? Both are things that satisfy the disjunctive property, black or not a raven. In the balance of this essay I want to consider briefly three approaches to solving the raven paradox, approaches to explaining why, as Goodman put it, there is no prospect for indoor

---

[6] C. Hempel, *Aspects of Scientific Explanation* (New York: Free Press, 1965), 14–20.

ornithology.[7] Why do some conjuncts confirm while others do not? The first two of the approaches I will consider are familiar, and in my view at once enlightening and problematic; the third is a sketch of one alternative, inspired by reliabilist approaches in epistemology.

Bayesians: Differential Priors

One popular solution to the raven paradox is Bayesian.[8] Bayesians hold that belief is a matter of degree and can be represented in terms of probabilities. Thus p(E) is the probability I give to the statement E, which may range from 0, if I am certain E is false, to 1, if I am certain E is true. By representing beliefs as probabilities, it is possible to use the mathematical theory of probability to give an account of the dynamics of belief, and in particular an account of inductive confirmation. The natural thought is that evidence E supports hypothesis H just in case the discovery of E causes me to raise my degree of belief in H. To put the point in terms of probabilities, E supports H just in case the probability of H after E is known is higher than the probability of H beforehand. In the jargon, what is required is that the posterior probability of H be greater than its prior probability.

What makes Bayesianism exciting is that the standard axioms of probability theory yield an equation that appears to tell us just when this condition of confirmation is satisfied, and so to give us a precise theory of induction. That equation is Bayes's theorem, which in its near simplest form looks like this:

$$p(H|E) = p(E|H)p(H)/p(E)$$

On the left hand side, we have the conditional probability of H given E. Bayesians treat this as the posterior probability of H, so the figure on the left hand side represents the degree of belief you should have after the evidence E is in. The right hand side contains three probabilities, which together determine the posterior. The first of these—$p(E|H)$—is the probability of E given H, known as the 'likelihood', because it represents how likely H would make E. The other two probabilities on the right hand side—$p(H)$ and

---

[7] N. Goodman, *Fact, Fiction and Forecast*, 4th ed. (Indianapolis: Bobbs-Merrill, 1983), 71.

[8] Cf. C. Howson & P. Urbach, *Scientific Reasoning: The Bayesian Approach* (La Salle: Open Court, 1989).

p(E)—are the priors of H and E respectively. They represent degree of belief in hypothesis H before the evidence described by E is in, and degree of belief in E itself before the relevant observation is made. This process of moving from prior probabilities and likelihood to posterior probability by moving from right to left in Bayes's theorem is known as conditionalising and is claimed by the Bayesian to characterise the dynamic of degrees of belief and so the structure of inference.

To get the flavour of the way Bayes's theorem works, consider a hypothetico-deductive case; that is, a case where H deductively entails E. Here the likelihood p(H|E) is simply 1, since if H entails E then E must be true given H. Under this happy circumstance, Bayes's theorem becomes even simpler: the posterior of H is simply the prior of hypothesis divided by the prior of the evidence: p(H)/p(E). Bayes's theorem thus tells us that a successful prediction of H confirms H, so long as the correctness of the prediction was neither certainly true or false before it was checked. For so long as the prior of E is less 1 but more than 0, p(H)/p(E) must be greater than p(H), and this is to say that the posterior of H will be greater than the prior of H, which is just when the Bayesian claims there is inductive confirmation.

Bayesianism seems at first to succumb to the raven paradox rather than solving it. For as the raven hypothesis entails that the next thing I see will be either a raven or not black, and this claim has a prior probability less than unity, the Bayesian is committed to the claim that whatever I see next will count as confirming the raven hypothesis so long as it does not contradict it, whether it be black raven, green emerald, or white paper. And indeed this is what the Bayesian says. But since confirmation on the Bayesian scheme is a matter of degree, it is open to the Bayesian to maintain that while most observations compatible with the raven hypothesis would only confirm it to a negligible degree, observing a black raven provides substantial confirmation.

Not only can the Bayesians this, but they have a good reason to say it. For the difference in confirmation here comes down to a difference in the priors of the evidence—the lower the prior, the greater the confirmation—and there is good reason to claim that the probability that the next raven I see will be black is substantially lower than the probability that the next non-black thing will be a non-raven. To see why this is so, notice that if there is a counterexample to the raven hypothesis, it will be a non-black raven, and so will be a member both of the class of ravens and of the class of non-black things. But given how many fewer ravens there are than non-black things, the probability of finding this counterexample by looking in the class of ravens is much greater than the probability of finding it by looking in the class of non-black things. The probability that a member of the class of ravens will be black will thus be lower than the probability that a member of the class of non-black things will be a non-raven, so observing a black raven confirms more than observing a non-black, non-raven.

Bayesianism is a powerful approach to describing our inductive practices, and as we see it has something enlightening to say about the raven paradox in particular. But I have three reservations about this solution. The first is that Bayesianism has been confronted with various general difficulties, not all of which it can clearly meet. To mention what is perhaps the best known of these, it faces the problem of 'old evidence'.[9] On the Bayesian scheme, what powers the increase from the prior to the posterior probability of H is a transition in the probability of the evidence from its prior value to its value after observation (effectively unity). But this contrast is only generated on the standard Bayesian scheme if H is formulated before E is observed. Otherwise, by the time H arrives on the scene its prior already reflects the impact of the earlier observation of E, so we have no before/after observation contrast. Thus without some special jiggling of the Bayesian scheme, a black raven observed before the scientist generates the raven hypothesis provides even less confirmation for that hypothesis than the subsequent observation of a piece of white paper, since it provides no confirmation at all.

A second reservation concerns the challenge with which I began this essay, the challenge of distinguishing genuine confirmation from mere content cutting. It is not clear that Bayesianism adequately illuminates that distinction. This can be seen in the context of the new riddle of induction. Recall the arbitrarily conjunctive hypothesis 'All emeravens are grack'. The Bayesian scheme entails that the observation of green emeralds does confirm this hypothesis, which seems the wrong result. In response to this, the Bayesian may say that the trouble with these gruesome hypotheses is that they have themselves such a low prior probability that even allowing that they are confirmed by their instances does

[9] Cf. J. Earman, *Bayes or Bust? A Critical Examination of Bayesian Confirmation Theory* (Cambridge, MA: MIT Press, 1992), chapter 5.

not lead them to end up with high probability.[10] But this is doubly wrong. First of all, even in a case where the posterior probability remains low, it is still seems wrong to say that there is confirmation. Secondly, as the example of 'All emeravens are grack' shows, gruesome hypotheses may have a high prior. To alter the example, consider a hypothesis with a small number of instances, such as 'everyone in the room is left-handed'. This may have a reasonably high prior probability, if there are few enough people in the room, and the Bayesian scheme will say it is confirmed by each person in the room we then find to be left-handed, yet this is a model of mere-content cutting, since the probability that the remaining people are left-handed does not go up.

I do not say that there is no way for the Bayesian to clarify the distinction between content cutting and genuine confirmation. For example, while the Bayesian notion of hypothesis confirmation is in itself only a kind of content cutting, presumably the Bayesian can also avail herself of various conditional probabilities that will make the distinctions we want. Thus I may assign higher probability to the claim that unobserved ravens will be black if observed ravens are black than to the claim that unobserved ravens will be black if observed emeralds are green. But this takes me to my third reservation. Even if the Bayesians can find a way of describing a scientist's web of probability assignments so as to reflect the inductive distinctions she makes, it is not clear how much light is shed on the sources of these degrees of belief. Bayes's theorem may be well seen as a kind of consistency constraint, the probabilistic analogy to a requirement that we not contradict ourselves.[11] Here the prior probabilities and the likelhoods are like the premises of a deductive argument, and the application of Bayes's theorem like the application of valid rules of deduction. But just as a consistency constraint does not say much about the source of the premises, so the application of Bayes's theorem does not tell us much about where the priors come from. Bayesians may be happy at this stage to maintain that subjective probabilities are just that; but it is I think worth investigating whether there are not any additional and more objective constraint on confirmation that our inductive judgements attempt to track.

---

[10] C. Howson, *Hume's Problem: Induction and the Justification of Belief* (Oxford: Oxford University Press, 2000), 239–240.

[11] C. Howson, *Hume's Problem: Induction and the Justification of Belief* (Oxford: Oxford University Press, 2000), chapter 7.

Peter Lipton

Quine: Unprojectible Complements

As Hume remarked in the quotation I have used as the epigraph for this essay, our inductive practices seem based on the presumption of a connection between the instance we observed and the instance we infer. The raven paradox shows that not only is the presumption difficult to justify, it is difficult even to acknowledge. The Bayesian analysis of induction, for all its merits, does not seem to do very well on this score. So I turn now to another well-known solution that may be more promising. This is the invention of Willard van Orman Quine, presented in brief compass at the start of his essay on natural kinds.[12]

Quine proposes an assimilation of the raven paradox to the new riddle of induction. Like many, including Goodman himself, Quine takes it that the new riddle shows that only certain predicates are suitable for induction. 'Emerald', 'green', 'raven', 'black' and their ilk are 'projectible'; but 'grue', 'emeraven' and 'grack' are not. Quine's clever proposal is that the complement of a projectible predicate is not projectible: thus 'non-black' and 'non-raven' are unprojectible, just like 'grue'. So just as the observation that past emeralds have been grue gives no reason to believe that future emeralds will be grue, so the fact that observed non-black things have been non-ravens gives no reason to believe that unobserved non-black things will be non-ravens. As Quine observes, this proposal can be made compatible with the equivalence condition on confirmation, if we say that black ravens confirm 'All ravens are black' and all its logical equivalents, and that white paper confirms none of them.

Quine's solution is appealing in part because it suggests a notion of connection behind the distinction we draw between cases of genuine confirmation and arbitrary conjunctions. Ravens tell us about other ravens, because ravens form a natural kind; non-black things do not tell us about other non-black things, because non-black things form a heterogeneous class, a rabble not a kind. But of course there are difficulties with this proposal too.

The raven paradox is not a puzzle about birds. To solve it, it is not enough to find a way to discount the epistemic power of non-black, non-ravens: the solution must generalise. Quine of course recognises this, which is why he suggests that the complement of every projectible predicate might be unprojectible.

[12] W. V. Quine, 'Natural Kinds', in *Ontological Relativity* (New York: Columbia University Press, 1969), 114–116.

But there are a number of difficulties with this suggestion. First of all, though this may not be a problem, it is worth noting that this proposal will not give a general account of unprojectibility, since many unprojectible predicates, such as 'grue', do not have projectible complements either. Not all the complements of unprojectible predicates (e.g. 'not grue') are projectible. More problematically, it is not clear that the all complements of projectible predicates are unprojectible either. Consider for example the hypothesis that all massless particles move at the speed of light. This, I take it, contains only projectible predicates. But I would say the same about its contrapositive that everything that does not move at the speed of light has mass. Or take the hypothesis that all invertebrates lack kidneys: this seems no more or less instance confirmable than its contrapositive.

A different sort of difficulty for Quine's proposal is that projectibility of predicates is not the whole story about the confirmability of hypotheses, since the same predicates may appear in both confirmable and unconfirmable hypotheses. Thus hypotheses such as 'All gold spheres have a diameter of less that one hundred miles' is a standard sort of example of a merely accidental truth not confirmed by its instances (though in this case we have excellent independent reasons for thinking it true, given our knowledge of the total amount of gold), yet there is nothing wrong with the predicates it contains to prevent them from figuring in other, instance confirmable hypotheses. Again, we may have inductive evidence for highly accidental generalisations. Thus if we sample randomly from a room of students, the observation that everyone in the sample is left handed does confirm the hypothesis that the others are left-handed too. Here what counts seems to be the method of sampling rather than the status of the predicates.

Like many philosophical problems, the raven paradox initially presents itself as a case where we know what the right answer ought to be, even if we cannot figure out how to arrive at it. Surely we know that other people do have minds, even if we do quite see how this knowledge is possible in the face of the sceptical problem of other minds. Similarly, it seems obvious that black ravens confirm the raven hypothesis and white paper does not; the trouble is in coming up with a description of our inductive practices which entails that result. But in fact it is not so obvious what the right answer is in this case. The paradox gains its initial power because we are inclined to say both that black ravens confirm 'All ravens are black' and that non-black non-ravens confirm 'All non-black things are non-ravens': we are inclined to agree with Nicod that hypotheses are confirmed by the observation of their instances. We are then discombobulated by the observation that these two hypotheses are logically equivalent. Thus the temptation to argue that the two hypotheses, correctly construed, are not really equivalent, so one can hold on to the judgement that each kind of evidence is confirmatory of its proper hypothesis. But the Bayesians, as we have seen, end up with a different result, according to which the right answer is that non-black, non-ravens provide only negligible confirmation for *either* hypothesis. And Quine goes even further, claiming on different grounds that they provide no confirmation at all.

So it is not so clear what the right answer is in this case, and I am not convinced that the answer Quine reaches is the one we should wish, since in fact the demarcation between confirming and non-confirming instances does not appear to follow the distinction between positive and contrapositive instances. The example of mass and speed is perhaps a case in point. Another and more pervasive illustration is given by a central feature of our inductive practices, dubbed by John Stuart Mill 'the method of difference'.[13] What Mill observed is that we very frequently make inferences from effects to causes on the basis of a *pair* of instances, one positive, one contrapositive. Thus in a controlled experiment the evidence that a certain drug lowers cholesterol is based on the combination of positive instances of people taking the drug whose cholesterol goes down with contrapositive instances of people whose cholesterol does not go down and who are not taking the drug but only a placebo. The inference depends on the appropriate pairing of instances, and has much greater force than either kind of instance on its own. In everyday life, application of something like the method of difference is ubiquitous, for example when we manipulate a cause to determine its effects. So an answer to the raven paradox that would rule out all contrapositive instances is probably not the answer we want.

My final reservation about Quine's solution parallels my concern that the Bayesian has not adequately illuminated the basis of judgements of confirmability. Quine's appeal to natural kinds seems more promising in this regard, because of the intuition that instances that are members of the same kind bear a deep similarity, and that this underpins our willingness to infer from what we observe in one member of the kind to others. But there are

[13] J. S. Mill, *A System of Logic*, 8th ed., London: Longmans, Green and Co., 1904) section III.VIII.2.

metaphysical difficulties in saying in what this similarity might be taken to consist. Moreover, it is not even clear how some metaphysically robust notion of natural kinds would help in this context. Grant that ravens are fundamentally like each other in a way that non-ravens are not. It is not clear why exactly this should matter when it comes to confirmation, given the logical equivalence with which the raven paradox began, and given that if ravens are united in what property they have, then non-ravens must be united in what property they lack.

Reliabilism: The Tracking Condition

An alternative approach to the Raven paradox that I now sketch appeals to a reliabilist epistemology which locates warrant in reliable belief-forming mechanisms. In contrast with an internalist epistemology that would make justification necessary for knowledge, where justification is a matter of reasons and argument, reliabilism is an externalist epistemology according to which what makes some true beliefs into knowledge is that the cognitive mechanism that gave rise to those beliefs was in fact reliable, not that the knower has reasons for thinking it so. Perceptual knowledge provides a good illustration of the reliabilist perspective. I know that there is a mug on my desk, not because I have made a warranted inference from a muggish experience to the existence of a mug in the world, nor because I have some general justification for relying on the testimony of my senses, but because I believe there is a mug on my desk, there is, and the perceptual mechanism by which the belief was acquired was reliable. Not to put too fine a point on it, my eyes are at the centre of a good mug-detection system, and that is what matters.

Reliabilism raises many questions and has been much discussed. I must ignore almost all of this in the present brief advertisement, but there are two issues I need to flag. One concerns the way the notion of reliability is to be understood. What does it mean to say that a belief-forming mechanism is reliable? One might for example try to analyse this in terms of a tendency or of a long-run frequency of truth generation, but here I will instead follow Fred Dretske and Robert Nozick and understand reliability as applying

Peter Lipton

to the single case, in terms of a counterfactual.[14] In its simplest form, the idea is that a true belief that p was reliability generated just in case had p not been true, it would not have been believed. Thus my belief that there is a mug on the table counts as reliably generated because, had there not been a mug on the table, I would not have believed that there was. To use Nozick's term, a belief that satisfies this sort of counterfactual 'tracks' the truth.

Another point about reliabilism that needs flagging is that while this form of externalism allows for knowledge without reasons or inference, it also applies to inferential knowledge. Inferential practices are themselves belief-forming mechanisms, and reliability provides a way of thinking about what makes something into a warrant-conferring inference. Reliabilism thus also provides a way of thinking about what makes something good evidence. What we want is an evidential policy that will lead us to accept hypotheses that are true, and where those true beliefs satisfy the tracking condition. We want to turn ourselves into reliable detectors of the truth value of the hypothesis in question: evidence helps us to do that.

We want a general epistemic policy that tends to take us towards truths and away from falsehoods. And I am suggesting that we also want our true beliefs to track. When we believe a true hypothesis, we also want it to be the case that we would not have believed it unless it were true. So when we believe H true because of evidence E, what we want to be the case is that, if H had not been true, we would not have inferred H from E. In one sort of simple case, this would amount to the condition that if H had not been true, E would not have been true either.[15] It is particularly easy to see how this is supposed to work in the case of an inference from symptom to underlying condition. Koplic spots are good evidence that someone has measles, because it is easy to determine that someone has koplic spots, because someone who has koplic spots is likely to have measles and, to come to the tracking condition, if that person did not have measles, she would not have had koplic spots either. And in the case where the hypothesis in question is a universal conditional, one way of construing a tracking condition is that our evidential situation be such that, had the hypothesis been false, we

---

[14] F. Dretske, 'Conclusive Reasons', *Australasian Journal of Philosophy* **67** (1971), 1–22; R. Nozick, *Philosophical Explanations* (Cambridge, MA: Harvard University Press, 1981), 172–196.

[15] R. Nozick, *Philosophical Explanations* (Cambridge, MA: Harvard University Press, 1981), 248.

would have seen a counterexample. A slightly different form the condition might take that has certain advantages for our purposes is that, had there been an unobserved counterexample to the hypothesis, there also would have been an observed counterexample. Thus where our observations give us good reason to believe that all ravens are black, it should be the case that had there been unobserved non-black ravens, there also would have been observed non-black ravens.

When our belief in a hypothesis satisfies the tracking condition, we have gotten ourselves into a situation where we would not have believed that hypothesis if had been false. Our evidential practices supported this by making it the case that we have looked in a place where H would have failed, had it been false. From this perspective, good evidence practice contributes to tracking, or makes tracking more likely.[16] My suggestion is that tracking is a confirmational virtue. This is not to say that tracking is the whole story of confirmation. For one thing, the sort of condition I have indicated only applies to true hypotheses, and we also need to make sense of the confirmation of hypotheses that are false. Good empirical technique should lead us to reject hypotheses that are false as well as lead us to establish a tracking relationship with hypotheses that are true. Nor would I claim that data that do not abet tracking must therefore be confirmationally irrelevant: they may have other epistemic benefits.

Nevertheless, the tracking condition may help to illuminate the distinction between genuine confirmation and mere content cutting, because it provides a connection between the observed and the unobserved. This is illustrated by the measles example: there is a counterfactual connection between the observed symptom and the unobserved disease. We can also see this connection in the context of a general hypothesis. Suppose that we truly believe a general hypothesis on the basis of observing some of its instances. The tracking reliabilist will say that still we do not know that the hypothesis is true if, had it been false, we still would have observed only positive instances. What is required for knowledge is that if the hypothesis had been false, we would have noticed, because we would have observed a counterexample.

Let us now apply this tracking reliabilist perspective directly to the raven paradox. As we have noted, all the objects in the world are

---

[16] There is more work to be done to articulate how this probability is to be understood, but I would like to capture the idea that confirmation in its tracking element is a matter of degree.

in a sense instances of the hypothesis, and will have the disjunctive property of being either black or a non-raven if the hypothesis is true. But it does not follow that every such object contributes to tracking, or contributes to tracking to the same extent. So far as the tracking condition is concerned, those instances that confirm do so to the extent that they would have been likely to be counterexamples if the hypothesis had been false. So the question to ask is this: What objects in the world would have been non-black ravens if not all ravens were black? I suggest that these objects would lie among the ravens, and not among the pieces of paper. A white piece of paper would not have been a non-black raven, had not all ravens been black, whereas a black raven might have been. Roughly speaking, the idea is that in the closest possible world where the raven hypothesis is false, actual black ravens are more likely to be non-black ravens than actual non-black non-ravens are likely to be non-black ravens.[17] And if this is right, it is not just because there are so many more ravens than there are non-black things that black ravens carry more evidential punch than white pieces of paper. Consider the hypothesis that all the ravens in a room (or in a possible world) are black, and assume that there are the same number of ravens in the room as there are non-black things in the room. Still, I would say that black ravens confirm the hypothesis and brown chairs do not. And I'm suggesting this is linked to the fact that it is the ravens rather than the non-black things that would have been non-black ravens, had there been any.

The search for evidence that would track is related to the Popperian idea that the scientist should look where a hypothesis is most likely to fail if it is false, but the conditions are not the same. For example, if we already know that some objects are black, we know they cannot be counterexamples to the raven hypothesis, so they would be of no further interest to the Popperian. But some of these black things (i.e. the ravens among them) may well contribute to tracking. So the tracking condition helps to account for the fact that it is sometimes worth gathering evidence even when it is known that the evidence will not contradict the hypothesis. Conversely, the observation of potential falsifiers, such as things antecedently known to be non-black which turn out on inspection also to be non-ravens, may not help much with tracking.

On the tracking view of confirmation, the reason that one conjunct of an arbitrary conjunction does not confirm the other is

---

[17] This way of putting the point was given to me by Paul Dicken.

that there is no tracking relationship between them.[18] The fact that past emeralds are green gives no reason to believe that future ravens are black, because even if those future ravens had not been black, the past emeralds would have been green. So even positive instances of some hypotheses—such as 'All emeravens are grack'—do not confirm them. This applies also to less contrived cases. Thus the tracking condition may help to explain why repeating the same experiment over and over again eventually yields diminishing returns, confirmationally speaking. At first the experiment improves the prospects of tracking the truth of the hypothesis being tested, but repetition beyond a certain point yields diminishing returns, because the later repetitions do not make tracking more likely. By contrast, variety in evidence—a traditional evidential virtue—does improve the prospects for tracking. (Various other standard evidential virtues, such as number and precision, may also have a tracking aspect.)

The picture of the tracking condition for confirmation that I have sketched so far is primarily of some instances tracking other instances: observing black ravens confirms the raven hypothesis because if some unobserved ravens had been non-black, then some of the observed ravens would have been non-black too. But the tracking condition has a broader application. For example, it helps us to see how observations that are not intuitively part of an hypothesis may nevertheless confirm it. It is not that only similar instances track each other: inferences from effects to causes may be particularly good trackers, as the reliabilist approach to symptom and disease suggests. The tracking condition may also fit well with the idea that evidence confirms hypotheses that would explain it well.[19] A good explanation makes a difference to the phenomenon being explained, which is to say that if the explanatory facts had been otherwise, the phenomenon would not have occurred. And this is tantamount to the condition that, had the explanatory

[18] As Tim Williamson pointed out to me, this shows why it is not fully adequate to construe the tracking condition as saying simply that if the hypothesis had been false the evidence would have been different. For suppose that the hypothesis is an arbitrary conjunction of the observed and the unobserved, where the observed conjunct is more 'fragile' or contingent than the unobserved conjunct. In this case the simple tracking condition would be satisfied even though there is no confirmation. What is required I think is that the observed track the unobserved, which is not the case for this conjunction.

[19] Cf. P. Lipton, *Inference to the Best Explanation* 2nd edition (London: Routledge, 2004).

hypothesis been false, the evidence would have been different. And the tracking condition may help to illuminate the various forms of hedged realism, where it is argued that one should not believe the full content of a high-level scientific theory. For it may be that the available evidence only helps to establish a tracking relationship with a part of the full theory: it is only if that part had been false that we would have noticed.[20]

The tracking condition is applicable beyond the context of instantial confirmation in yet further ways. For the probability of tracking depends not just on the content of the evidence, but also on facts about how it was gathered. Take the case of the over-zealous lab assistant. Knowing how ego-invested I am in my pet hypothesis that all ravens are black, my assistant adopts a policy of only showing me ravens that are black, a policy of quietly suppressing any counterexamples he should come across. As it happens, he never has to implement his strategy, since all ravens are in fact black; nevertheless, the black ravens I am shown do not provide strong evidence that all ravens are black since, in part because even if there had been counterexamples, I never would have seen them. On the positive side, the tracking condition may help to account for the fact that a policy of random sampling makes it possible to confirm certain hypotheses that are 'accidental'. Thus polls may give good evidence about the distribution of views on some topic even though the hypothesis that the views have that distribution is not particularly lawlike, because the sampling strategy satisfied the condition that if that views had been differently distributed, the poll results would have been different too.[21]

How does this tracking perspective on confirmation compare with the Bayesian and the Quinean approaches to the raven paradox? Unlike the Bayesian, the tracker appeals to a genuine connection between instances, not just content cutting. And where Bayesian probabilities are subjective, tracking relations are objective, or anyway more objective than brute degrees of belief. And where the Bayesian looks for evidence with a low prior probability, the tracker looks for what will improve the prospects for tracking, and these are different features. On the one side there can be evidence that does not have low prior but tracks, such as old evidence that tracks. On the other side there is also evidence that

[20]   Cf. A. Chakravartty, 'Semirealism', *Studies in History and Philosophy of Science* **29** (1998), 391–408.

[21]   Cf. J. Woodward, 'Realism about Laws', *Erkenntnis* **36** (1992), 200.

has a low prior but does not track, such as an arbitrary and unlikely conjunct. And for the tracker, when it comes to classes, size isn't everything: members of the larger class may contribute to tracking and members of the smaller class may not.

Unlike Quine, for the tracker what is important is not the predicate per se, but whether one instance tracks another. But it may be that natural kinds are precisely classes where members do tend to track each other. And perhaps we may say that a predicate is projectible when one instance tracks another. But the tracker can also make sense of the confirmational value of contrapositive instances, for example in the context of Mill's method of difference. One of the reasons that the positive-contrapositive pairs of instances in a controlled experiment can provide such strong support for a causal hypothesis is because well-chosen pairs of this sort are very likely to track. Thus in a well-controlled experiment to test whether a drug lowers cholesterol, where the outcome is positive, we would not have gotten those results if the drug did not really have that effect. Either the cholesterol of the drug takers would not have decreased, or the cholesterol of the placebo takers would have decreased. The combination of well paired positive and contrapositive instances maximize the chances of tracking. This also helps to explain why what makes for good pairing in the method of difference depends on relevant similarity between the members. We don't want the people in the control group to have a very different diet or a very different level of physical activity than those in the group that gets the drug. If the drug takers also take unusual amounts of exercise and have an exceptionally healthy diet, we might well have gotten the contrasting experimental results even if the drug had not lowered cholesterol, and that is one reason why such an experiment would not provide good evidence.

Scientists want to get themselves into the happy cognitive situation where they know that certain hypotheses are true. To do this, they must find hypotheses that are indeed true, but that is not enough. They must also hook up with those hypotheses in a tracking relationship, and this is something that good evidence and good epistemic practices enable them to do. As I have suggested, to see how they do this, suppose that the hypothesis in question is true. Now consider how the world would have been different if the hypothesis had been false. Scientists are themselves part of the world. In order to know the hypothesis to be true, scientists need to get themselves into a situation such that in the nearby world where the hypothesis is false, it is also the case that they do not believe it to be true. Certain observations and practices help them to achieve this; others do not. Observing black ravens may help, if the hypothesis in question is the raven hypothesis; but this is not simply because black ravens are instances of the hypothesis. And pieces of white paper may not help, even though they are instances of the contrapositive. What matters is what would have been different and what the scientist would have noticed.

Conclusion

This has been a lightning sketch of how one might use the tracking approach to illuminate aspects of empirical confirmation in general and how it might help with the raven paradox in particular. But I do not suffer from the illusion that tracking is the whole story about confirmation, or that the account will work for all cases in the rough form here presented. There are also challenging questions about the epistemology of the counterfactuals on which the account depends. How do we judge whether a tracking relation is likely to hold? And there are other tricky features of counterfactuals that raise challenges that a properly developed tracking account would have to meet. For example, since evidence is in the past and prediction is in the future, the tracking account requires that we allow counterfactuals to backtrack, so that it may be true to say that if were to have been different in the future, things would also have been different in the past.[22] A different sort of worry arises from the possibility that some empirical hypotheses are necessarily true, since in that case it appears we cannot consider what would have been the case had such a hypothesis been false.[23]

Nevertheless, appeal to the tracking condition seems to me a promising approach to confirmation and to the raven paradox because, unlike other well-known approaches, it does aver to a link between the observed and the unobserved. And here it seems to me that Hume was right: insofar as we take the observed as a guide to the unobserved, this is because we suppose that there is some connection that binds them together. From a tracking perspective, the solution to both the New Riddle of induction and the Raven Paradox depends on seeing that only some instances track others.

[22] Cf. D. Lewis, 'Counterfactual Dependence and Time's Arrow', in *Philosophical Papers*, Vol. II (New York: Oxford University Press, 1986), 32–66 .

[23] Cf. S. Kripke, *Naming and Necessity* (Oxford: Blackwell, 1980), Lecture III.

And the difference between an arbitrary conjunction and a confirmable conjunction is that only in the confirmable case do some conjuncts track others.

The tracking condition has its origin in an account of knowledge rather than in an account of justification or confirmation. So insofar as this condition does indeed have a role to play in confirmation, perhaps we should see the aim of empirical testing of scientific hypotheses not just as the discovery of the truth, but also as the acquisition of knowledge. On this view, scientists seek through their empirical work to hook themselves up with nature in a certain counterfactual way. At the start of this essay, I distinguished between the justificatory and descriptive tasks in epistemology. I think the appeal to tracking is part of both. It is part of the descriptive task, where the raven paradox naturally appears, because I have suggested that we can account for our differential attitude towards the relevance of black ravens and white paper in part in terms of our judgement of a differential contribution to establishing a tracking relationship between evidence and hypothesis. But an appeal to tracking also forms part of a justificatory or normative account, since it is part of an account of what knowing that a hypothesis is true requires. So an appeal to a tracking condition has a place in an account of what it takes to have scientific knowledge, and we can improve our description of scientific practice by seeing inquirers as attempting to meet that condition.[24]

University of Cambridge

[24] I am grateful to audiences at Oxford University, Cambridge University, University of East Anglia, Leeds University, University of Aarhus, and the Royal Institute of Philosophy, and for helpful reactions to my talks about ravens, and to Alex Broadbent and Paul Dicken for comments on a draft paper.

What's the point in Scientific Realism if we don't know what's really there?

SOPHIE R. ALLEN

The aim of this paper will be to show that certain strongly realist forms of scientific realism are either misguided or misnamed. I will argue that, in the case of a range of robustly realist formulations of scientific realism, the 'scientific' and the 'realism' are in significant philosophical and methodological conflict with each other; in particular, that there is a tension between the actual subject matter and methods of science on the one hand, and the realists' metaphysical claims about which categories of entities the world contains on the other.

First, I will clarify the definition of scientific realism, since both its proponents and its critics tend to characterize it in several apparently different ways and it will be instructive to collect together those views which will be covered by what I have to say. Some well-known views which are also called 'realist' will be largely immune to my attack and I will have a little more to say about these at the end of the paper when I consider whether any weaker forms of scientific realism are tenable in view of the objection I have raised. In the second section, I will outline the strategy with which I intend to criticize scientific realism and distinguish it from other, more commonly encountered, objections to the scientific realist view. In the third section, I will apply this strategy to consider two examples of ways in which current scientific thought is at odds with generally accepted accounts of categories of entities or processes in metaphysics. I will argue that such considerations prompt scepticism about our ability to characterize the entities and processes we expect our science to discover, either a priori or after extensive scientific investigation, except in terms so general that our characterizations are in danger of becoming vacuous and the distinctions between categories eroded. In view of this, the very definition of scientific realism in its more robust realist formulations is threatened with either being straightforwardly false, or a trivially true, and thereby vacuous, doctrine which falls far short of what most scientific realists intend.

Sophie R. Allen

All of this is of no import to those who do not espouse a robust form of realism about the postulates of our best, or at least better, *current* scientific theories; but those who do are left with a considerable problem concerning how the doctrine of scientific realism can retain any content at all. At best, scientific realists seem to be held hostage to the universal success of some hypothetical future science—the one which explains everything and perhaps predicts it as well—which might be immune to objections of the type I will raise. To assess the potential of this response, I will consider the extent to which it is plausible for the scientific realist to 'wait and see' where science leads and to take a robustly realist attitude to the entities postulated by this hypothetical, future, ideal theory. My inclination is to be pessimistic about the vindication of robust scientific realism in this matter too, since this strategy can only deliver realism if there can only be one such theory, or a very narrow range of theories,[1] which are completely explanatory as required. Otherwise, faced with several empirically equivalent contenders for the role, the resultant doctrine would be significantly weaker than the one which the robust scientific realists hoped to defend (although it is sometimes honoured with the name 'realism'). This conclusion will depend on the resolution of some issues which lie outside the scope of this paper, however; so I will not be able to definitively defend it here.

So, what conclusions can be drawn about scientific realism in the face of the difficulties I have raised? I will suggest that since, in its robust form, it cannot reconcile its commitments to realism and to science, it is a doctrine which is difficult to defend and that alternative accounts of what our science is about should be explored.

1. Some Definitions

Roughly speaking, by 'scientific realism' I mean the collection of views which maintain that we should take a realist view about the

---

[1] The set of entities which appear in the future theory only needs to be 'unique-ish' rather than unique in order to permit realism in some strong form, since even the most committed realists are prepared to admit that science can only specify what there is within a reasonable margin of error. The possibility of realism in the future would become untenable, however, were there to be a wide range of empirically equivalent theories which postulated radically different theoretical entities from each other.

entities named in our current scientific theories; in particular, that the entities named in scientific theories actually exist in the objective world, have the properties which empirical investigation tells us they do, and are neither artifacts nor postulates of our theorizing, nor are they otherwise dependent upon it. Theoretical entities exist and do so mind-independently. So, for instance, if our physics talks about—or, more formally, quantifies over—electrons, photons, gravitons, gluons and quarks, then such things exist in the objective world with the properties which our theory attributes to them: science tells us about the way reality *is*. Likewise, if our biology and psychology find that talk of genes is indispensable, then genes too exist in the manner in which our best genetic theory says they do.[2]

However, the term 'scientific realism' means different things to different people, and it is sometimes characterized in alternative ways. In addition to its being defined as a metaphysical or ontological doctrine about the existence of theoretical entities as I have above, it is also treated as being a semantic doctrine about truth or reference: the claim that our best scientific theories are true, or at least approximately true;[3] or, that the predicates of our theoretical language genuinely refer.[4] Thirdly, scientific realism is also defined in epistemic terms, to the effect that we are justified in believing that the entities which our science postulates exist mind-independently.[5] But despite the *prima facie* differences

---

[2] For some interesting scepticism about genetic theory, see the previous *Supplement* to *Philosophy*: *Philosophy, Biology and Life* (Cambridge: Cambridge University Press, 2005), *passim*, especially John Dupré's article 'Are there Genes?', 193–210.

[3] I will not pause to consider the difficulties involved in clarifying the concept of 'approximate truth'.

[4] See, for example: Richard N. Boyd, 'The Current State of Scientific Realism', in Jarrett Leplin (Ed.), *Scientific Realism* (Berkeley: University of California Press, 1984), 41–2; Evan Fales, 'How to be a Metaphysical Realist', in Peter A. French, Theodore E. Uehling and Howard K. Wettstein (Eds.), *Midwest Studies in Philosophy XII: Realism and Anti-Realism* (Minneapolis: University of Minnesota Press, 1988), 253–4; Richard Jennings, 'Scientific Quasi-Realism', *Mind* 98 (1989), 240; Carl Matheson, 'Is the Naturalist Really Naturally a Realist?', *Mind* 98 (April 1989); David Papineau, *Theory and Meaning* (Oxford: Clarendon Press, 1979), 126.

[5] See Andre Kukla, *Studies in Scientific Realism* (Oxford: Oxford University Press, 1998), 10; Stathis Psillos, *Scientific Realism: How Science Tracks Truth* (New York: Routledge, 1999), xix-xxi.

Sophie R. Allen

between these different formulations and the ontological one, they are not, in any interesting philosophical sense, specifically semantic, nor epistemic, respectively, since they do not involve commitment to any particular theory of truth, reference or knowledge. I shall therefore retain the ontological formulation for the purposes of this discussion.

However, even from its supporters' perspective, before any serious philosophical objections against it have been raised, the rough definition of scientific realism that I have given so far stands in fairly obvious conflict with the practice of current science and so, in order to retain some plausibility, scientific realism needs to be qualified in a number of ways. The ontological commitment which the scientific realist espouses is intended to apply to the entities picked out by our current scientific theories, but it would be highly implausible to suggest that our current science is infallible, that all current theories are equally well confirmed, or that every entity postulated by our most successful and well-confirmed theories is essential to it, with no entity ever being postulated or retained as matter of convenience. Consequently, in order to allow for such difficulties as the fallibility of empirical investigation, and the pragmatic inclusion of theoretical terms which serve to simplify the formulation of a theory, scientific realists are content to introduce a certain amount of vagueness into their definitions. For example, take Michael Devitt's recent definition of a strong form of the scientific realist doctrine as follows:

> Most of the essential unobservables of well-established current scientific theories exist mind-independently and mostly have the properties attributed to them by science.[6]

The removal of the final clause that real scientific entities 'mostly have the properties attributed to them by science' characterizes a weaker, existential version of scientific realism, which is nevertheless intended to contain implicit ontological commitment to every unobservable 'having whatever properties are essential to its nature as that unobservable'.[7] This thereby involves greater ontological commitment than a purely referential form of non-essentialist realism which might treat theoretical terms as purely referential designators; that is, as terms which refer to something-or-other,

[6] Michael Devitt, 'Scientific Realism' in Frank Jackson and Michael Smith (Eds.), *The Oxford Handbook of Contemporary Philosophy* (Oxford: Oxford University Press, 2005), p. 769.

[7] Ibid., 769.

regardless of what our science says about whatever they refer to, and regardless of whether distinct kind terms of our theoretical language refer to objectively individuated, mind-independent kinds.

It is worth noting here that the opponents of the scientific realism which I am targeting might not be opposed to realism in this latter sense, since they are not necessarily opponents of realism, *per se*; that is, they do not always—or even usually—count themselves as being sceptics about the existence of the external world, as idealists, phenomenalists or verificationists. Rather, their scepticism is rather more restricted in scope and concerns the existence, or the nature, of the types of entities which the theory postulates or, even more narrowly, what might be called the 'unobservables' postulated by scientific theories.[8] Such entities either do not exist, they claim, or they do not exist entirely mind-independently; that is, they do not exist independently of humans theorizing about them.

Finally, I do not mean to turn this paper into a question about epistemology: the 'know' of the title is meant to be understood in largely common-sense terms, without any particular account of knowledge in mind. Those who would prefer not to think in terms of knowledge at all may substitute 'justified belief', or their favoured alternative: our 'not knowing what's really there' is not intended to be trivially true due to a high standard being required for the attainment of knowledge.

2. The Strategy

I am by no means the first to question the tenability of this strong version of scientific realism and previous critics have already concocted some well-known challenges to it.[9] The realist claims that 'most of the essential unobservables of well-established current scientific theories exist mind-independently and mostly have the properties attributed to them by science' but these uses of

---

[8] I use the term 'unobservable' with some hesitation, since the distinction between what is observable and what is not is notoriously difficult to define. What I have to say in this paper will not rely on there being a principled distinction between observables and unobservables, nor on our being able to make it.

[9] See almost any introductory text on the philosophy of science for a review of the standard arguments.

'most', 'mostly', 'essential' and 'well-established' are fraught with philosophical dangers. In allowing that some theoretical entities in a theory are not essential to it, that some theoretical entities do not objectively exist, and that some of our current science is false, the realist is open to the charge that the arguments for these concessions can be generalized and the realist view is opened up to wholesale scepticism about all the entities a theory postulates; thus the tenability of realism is brought into question. For example, there are epistemic difficulties concerning how we could ever know which entities of a specific theory were essential to it: perhaps for every entity or set of entities in a theory T_1, an alternative empirically equivalent theory T_2 can be formulated which does not refer to that entity or set of entities, yet (being empirically equivalent) T_2 predicts and explains the same phenomena as T_1 does.[10] Similar difficulties arise in light of the realist's admission that at least some of our current scientific theories are not well enough established for a realist attitude to be adopted towards the entities they postulate: in addition to the obvious epistemological problem of determining which theories these are, the critic of realism points out—with the so-called 'meta-induction from past falsity'—that all our past scientific theories have turned out to be false, so it is fair to infer that our current scientific theories are probably false too, no matter how well established they are. Furthermore, past scientific theories have postulated entities which turned out to lack many of the properties attributed to them, such J. J. Thomson's 'plum-pudding model' of the atom,[11] or entities, such as the well-known philosophical example of phlogiston, which turned out not to exist after all; in this latter case, the problematic term 'phlogiston' turned out to refer to the absence of something, the absence of oxygen. But phlogiston theory, championed by Joseph Priestley, was once as well-established as some of our more respected current science, and for the latter half of the 18th

[10] Contrary to what some scientific realists seem to think (including Devitt, *op. cit.*), this objection does not require that the empirically equivalent rival theory T_2 does not refer to any unobservables at all, merely that it refers to some different ones from those postulated by T_1. This makes it more plausible that there are such empirically equivalent theories, although I will not pursue this matter here.

[11] Thomson's model of the atom treated electrons as being embedded in a cloud or soup of positive charge, like plums embedded in plum pudding, hence the name; his student Rutherford directed the 'gold foil experiment' which led to this model being abandoned in favour of a model of the atom as having a nucleus orbited by electrons.

Century there was an ongoing debate about whether a replacement to phlogiston theory was required.

Instead of evaluating the realists' responses to these well-worn objections, I intend to pursue a different strategy since there seems to be room for some scepticism of a rather different flavour concerning scientific realism in its robust form. By reflecting on debates arising in current science, I will suggest that the theoretical flexibility of scientific theorizing outruns the metaphysical flexibility of the the realist's ontology—his account of which categories of things there are in the world—and thus that scientific theorizing can lead us to be sceptical about the categories of entities which the robust realist takes to exist mind-independently. Such scepticism might be concentrated upon what such general species of entities are like and how they should be characterized, but also, more seriously, if we are forced to stray too far away from a recognizable characterization of their essential nature (or resort to a characterization which is too broad to retain enough content to distinguish the entities in question are from those of other categories), one might be forced to question whether we can or should usefully think of such a category of entities as existing at all. For instance, in contrast to the traditional problems concerning phlogiston and the like raised above, instead of past mistakes in the history of science prompting the question of whether the types or kinds classified by the predicates of our scientific theories are the ones there really are—that is, the ones which exist mind-independently—one might be led to ask, for example, whether there are any mind-independent types or kinds at all, if it appears that scientific practice permits too broad a conception of what such entities can be.

It is important for scientific realists that there are some such, reasonably fixed, characterizations of the categories or species of entities which they take the terms of our theories to be trying to pick out, in the first instance because some such understanding is presupposed by the definition of scientific realism itself.[12] Moreover, although scientific realists are bound to differ in their ontological commitments and to disagree about the exact nature of the categories of entities which exist, there is nevertheless a significant amount of agreement, since realists require entities to serve similar roles: one realist might prefer an ontology of tropes

---

[12] This need not go as far as the provision of constitutive identity and individuation criteria for the members of a category of entities, although it may do so.

(say), while another favours universals. However, these differences can be overlooked for the purposes of this discussion, since the problems I will raise will be applicable whichever specific category of entities is postulated by the realist as the ontological grounding of objective qualitative sameness and difference between spatio-temporally distinct entities; one cannot, for example, avoid the difficulties I will discuss by opting for a subtle ontological shift from universals to tropes.

Secondly, whether the entities in question be causal laws or processes, particles, properties or natural kinds, the common response to the objections to scientific realism which cite previous reference failures is to treat such errors in the same manner as perceptual mistakes, perhaps by denying that the scepticism of the realist's opponent is globally applicable. For instance, the realist can argue that the situation with regard to the non-existence of phlogiston was simply that scientists of the time (namely Priestley and his supporters) were mistaken about the existence of a certain type of stuff, the existence of which can be rejected by the realist in favour of postulating another type, or types, of stuff.[13] Simply because we have been mistaken about whether some of our theoretical terms have referents, that does not imply that none of them do and that our theories do not, in the main, pick out the correct types or kinds of things. Alternatively, the realist can widen the gap between metaphysics and epistemology by distinguishing the ontological claims he makes about the existence of theoretical entities from how we know about them. To this end, he can point out that, despite there being epistemic problems associated with determining exactly which theoretical entities postulated by our current science actually exist—that is, determining which of our current theoretical terms will go the way of phlogiston—the opponent of realism has not shown that such entities do not objectively exist.

If, as I will argue, science gives us reason to think that our characterization of fundamental ontology is insufficiently secure—that we do not know, in a broad metaphysical sense, what is really there—then the project of linking the terms of our theories with 'really existing' entities in the world seems to be rather premature. Moreover, if science gives us reason not to be robust scientific

---

[13] In fact, phlogiston was initially replaced by Lavoisier with the almost-as-mysterious *caloric*, an 'invisible' and 'weightless', 'imponder-able fluid' which was only later absorbed into thermodynamics by Joule's identification of heat with the kinetic energy of molecules.

realists, then the motivation for, and plausibility of, such robust scientific realism is significantly eroded.

There are certainly plenty of purely metaphysical concerns which can be raised about the existence or non-existence, and the nature, of entire metaphysical categories of entities which realist science takes for granted, but I intend to explore whether there are any empirical considerations which could present problems for the plausible characterization of a category of entities as a whole. I will present two such examples where scientific theorizing has out-run the, relatively-speaking more a priori, metaphysical theories, such that if the realist contents himself with the current metaphysical realist story about which categories of entities there are, then certain areas of science simply could not be true: there is a *prima facie* clash between the realist's naturalism, embodied in his endorsement of the scientific method, and his commitment to realism about the theoretical entities of our current science.

Some philosophers will immediately find this strategy problematic, since they will consider that it accords an unsuitable level of importance to scientific theorizing and empirical method in order to draw out implications about metaphysical matters which, they will object, should be treated as a priori.[14] Why, for instance, is the inclusion of empirical data in philosophical study not akin to the long-discredited scientism, which allows empirical investigation to take the place of a priori theorizing? It is important to note here, however, that the strategy I will pursue diverges from a problematic scientistic one of letting the findings of science dictate what we think that there is in the world. For the purposes of this debate, it does not matter whether the examples I present actually are true, merely that they could be (in quite a strong sense of physical possibility[15]); this is fortunate, since invoking a constraint on the examples actually being true would simply result in my having to beg the question that we know what the truth of a scientific theory consists in, over and above a certain amount of empirical success. Since the metaphysical picture which scientific realism implicitly presupposes is intended to cover all possible empirical outcomes—

---

[14] I have defended this strategy at greater length elsewhere. See 'Disorder at the Border: Realism, Science and the Defense of Naturalism', *Philo* 7, No. 2 (2004), 183–185.

[15] I take it here that the prospect that a theory could be empirically confirmed or disconfirmed is a reasonable indication of its hypotheses being physically possible, since the prospect of empirical confirmation indicates that such hypotheses may well actually be true.

the entities and processes which realism espouses are supposed to exist however empirical investigation eventually tells us the world is—our metaphysical reflections should be able to deal with a wide range of physical possibilities which might be true of the actual world. In particular, a metaphysician of a scientific realist mien ought to avoid ruling out scientific hypotheses a priori in virtue of his metaphysics such that some empirical hypotheses *could not be true* within his metaphysical worldview. A scientific realist metaphysics should therefore respect a principle of plenitude about the ways which the world may turn out to be on empirical investigation, permitting as wide a range of physical outcomes as can reasonably be formulated, and in this admittedly inexact process it seems both obvious and sensible to use scientific theories as a defeasible guide.

Of course, this strategy does presuppose some minimal form of philosophical naturalism which will not be to everyone's taste, since it treats ontological theorizing as being continuous with empirical theorizing, rather than their being the products of radically differing modes of thought. However, I think it is plausible to expect that scientific realists should already be committed to a minimal form of such naturalism, since the conjunction of a commitment to scientific realism with entirely aprioristic metaphysics has the potential to create considerable tension between what our best scientific theories say the world contains and the ontological commitments endorsed by prior metaphysical theorizing. Thus, the scientific realist who strictly limits himself to a priori metaphysics is immediately in danger of undermining the tenability of his own scientific realist position.

3. Current Science and Ontological Categories

Laws and Quantum Theory

The first example I will consider is the least speculative and represents something of a historical test case about how realist metaphysical commitments might change in the light of empirical science. It concerns the well-known case of how the presumption in favour of causality being deterministic suffered as a result of findings in quantum theory. Less than a century ago, a scientific realist who was committed to the definition above—that most of the essential theoretical entities of contemporary science exist, and have most of the properties which science attributes to them—

would have been committed to the objective, mind-independent existence of a category of processes generally known as deterministic causal laws. The mind-independent world partially constituted by such laws would have run rather like clockwork, such that each state of the universe was entirely and uniquely determined by past states, and everything that had happened or would happen could, in principle, have been predicted from the beginning of time. From the point of view of a scientific realist of that era, it would have been plausible to think that current science was wrong about some of the deterministic causal laws it postulated, but not that there were no such laws at all.[16]

However, quantum theory conflicts with this worldview of mind-independent reality being governed by deterministic causal laws, since according to that theory, a prior set of initial conditions does not uniquely determine a particular outcome, only the probability of that outcome occurring. The Schrödinger equation, which describes evolution or change in the quantum realm, describes behaviour deterministically but it does so in terms of the deterministic evolution of *probabilities* known as the 'wave-function'. The location of an electron, for example, can best be described in terms of the probability of its being found at a particular location, although when a measurement of location is made, the electron will be found at a particular location and it is no longer possible for it to be anywhere else, which prompted John von Neumann to postulate that the wave-function indeterministically and discontinuously collapses to a particle-like state.[17]

---

[16] It would be a mistake to think that scientists in the late 19th Century did not think their theory (their physics, at least) to be almost complete, or at least making excellent progress on the right lines with only a few loose ends remaining to be tidied up. (See, for example, Albert Michelson's 1894 speech, quoted in John Horgan, *The End of Science* (New York: Broadway Books, 1997), 19; or Lord Kelvin's 1900 address to the British Association for the Advancement of Science.) This optimism was shattered by Max Planck's proposed solution to the problem of black body radiation in 1900 which introduced quanta of energy and led to the formulation of quantum theory.

[17] There is widespread disagreement about whether von Neumann's postulated collapse of the wave function should be realistically construed, and if so, whether and how it is brought about by measuring quantum particles (known as The Measurement Problem), or if the collapse also occurs where no measurement is being taken. (See Nancy Cartwright, 'How the Measurement Problem is an Artefact of the Mathematics' in her *How the Laws of Physics Lie* (Oxford: Clarendon Press, 1983) for a version

Sophie R. Allen

One can choose to defend a deterministic view of causality in light of these probabilistic and indeterministic descriptions of the quantum realm, but it appears almost impossible to do so while maintaining a scientific realist perspective. Firstly, there is a popular defence of determinism which treats the entities of quantum theory as nothing over and above descriptions, taking an instrumentalist or anti-realist view of the probabilistic entities and processes they postulate, while maintaining a determinist and realist view of the entities in theories concerning the non-quantum realm.[18] However, as a defence of scientific realism, this approach is in danger of being completely ad hoc, since it is not obvious whether it is possible to draw a principled distinction between quantum and non-quantum theories in this way, treating the entities of the latter as mind-independent and the entities of the former as instrumental. It is a simple enough matter to construct thought experiments in which the latter affect the former—that is, the indeterministic processes of the quantum theory have effects on the ordinary middle-sized objects of our everyday experience—so the distinction cannot simply be drawn on the basis of size. Take, for instance, Feynman's example of a bomb that is set to go off when a Geiger counter registers a certain reading, or Schrödinger's famous cat[19], which make this selective approach seem evidently absurd: I will take it that even those who are prepared to defend realism about deterministic laws by selectively denying realism would prefer not to be instrumentalists or anti-realists about cats.

The second, alternative approach which is consistent with retaining realism about deterministic laws is to regard the conflict

---

of the latter view.) I will not go into the intricacies of this issue here, however, since there are difficulties enough for the notion of deterministic causality from the Schrödinger equation alone; should it turn out to be preferable to treat the indeterministic collapse of the wave function realistically, then the defence of determinism will be significantly more difficult than I suggest here.

[18] See, for example, Ted Honderich, *Mind and Brain. A Theory of Determinism*. Volume 1 (Oxford: Clarendon Press, 1988), sections 5.6–5.7; and *How Free Are You?* (Oxford: Oxford University Press, 1993), Chapter 6.

[19] See E. Schrödinger, 'Die gegenwärtige Situation in der Quantenmechanik', *Naturwissenschaften* 23 (1935). Translated into English by J. D. Trimmer: 'The Present Situation in Quantum Mechanics: A Translation of Schrödinger's 'Cat Paradox' Paper', *Proceedings of the American Philosophical Society* 124 (1980), 323–338.

between deterministic laws and quantum theory as being merely apparent, a sign of the inadequacy or incompleteness of our current theory which might ultimately be resolved by the discovery of a hidden variable, some as-yet-unnoticed feature of the world that would permit the formulation of a theory in which determinism is restored. Such a discovery would ensure that our science, realistically construed, did not deny the existence of an entire category of entities which we had, on the realistic construal of earlier theories, previously presumed to exist. However, for the scientific realist, this strategy has two significant drawbacks: the first being whether such a strategy is tenable in conjunction with the definition of robust scientific realism which puts its faith what our *current* theories say there is in the world; and the second, whether this response is a plausible approach to the problem at all. The former difficulty arises since, in order to put his faith in the restoration of determinism to quantum theory, the scientific realist is forced to admit that some of our current theories are simply wrong about the nature of an entire metaphysical category of entities, despite the fact that our current quantum theory is extremely successful, empirically speaking, as a predictive and explanatory tool. This admission could only lend ammunition to the critics of scientific realism in general though, since it reinforces the epistemic worries about which theories are to be trusted with our ontological commitment and which are not, a question which it seems the realist is finding all the more difficult to answer in a non ad hoc manner. (One cannot simply say, for instance, that the ontological commitments of quantum theory are not to be trusted, since the only basis for this seems to be that they do not fit the determinist realist's favoured metaphysical account of the world as containing deterministic laws.) This general difficulty is compounded, however, by the second drawback to this approach which suggests that it might not work at all since there are further well-confirmed empirical considerations which constrain the form such a theory could take.[20] In particular, if a version of quantum theory is to be devised which explains our empirical observations in

[20] These difficulties were initially formulated by J. S. Bell ('On the Einstein Podolsky Rosen Paradox', *Physics* 1 (No. 3), 195. Reprinted in his *Speakable and Unspeakable in Quantum Mechanics* (Cambridge: Cambridge University Press, 1987)) and gained some important empirical confirmation in Alain Aspect, Jean Dalibard, Gérard Roger, 'Experimental Test of Bell's Inequalities Using Time-Varying Analyzers', *Physical Review Letters* 49 (No. 25, 1982), 1804–1807.

terms of deterministic causal processes, then this is unlikely to be the kind of localized deterministic causality which is familiar from the usual philosophical examples of bricks breaking windows, or billiard balls colliding with each other. Rather, it is likely that any hidden variable would be a non-local phenomenon—that is, we are not simply searching for a hitherto unnoticed feature of the interacting entities under observation—and this embrace of a holistic account of causality or change might make it very difficult to maintain a realist commitment to other categories of entities such as discrete particular objects or events, or properties and lawlike relations between them. For example, David Bohm's account of quantum theory denies the existence of discrete particular entities, such as particular particles; for him, such entities should be regarded analogously to vortices in water which exist in some senses qua particular entities and yet are inseparable from the water in which they form and from each other.[21] Put in more general and simpler terms: the way such versions of quantum mechanics says the world is not the way that most of our other theories say it is, and so realism about some, or all, of the categories of entities in the latter would have to be abandoned. So a quantum theory which successfully avoids indeterminism and probabilistic causation, permitting the realist to maintain his commitment to deterministic causation, is unlikely to be one which is also compatible with maintaining realism about entities of other ontological categories to which scientific realists are generally committed, in some cases in virtue of the definition of their doctrine itself.

Although much work has been done to devise an account of quantum theory which sustains deterministic causation, the plausible candidates are few[22], and it would not be an exaggeration to say that the current consensus in physics is that there is very little likelihood that such a theory will be formulated. Contemporary scientific realists too are now much more likely to accept that they must allow for the existence of probabilistic laws or indeterministic processes in the mind-independent world, perhaps

[21] David Bohm, *Wholeness and the Implicate Order*, (London: Routledge and Kegan Paul, 1980), 10.
[22] Perhaps the most plausible interpretation of quantum mechanics which retains determinism is David Bohm's pilot wave theory. See Bohm, *ibid*. and D. Bohm and B. J. Hiley, *The Undivided Universe: An Ontological Interpretation of Quantum Theory* (London: Routledge & Kegan Paul, 1993).

at the most fundamental level: prompted by changes in physical theory, they are happy to disregard what would have been considered an essential feature of a whole category of processes which were a mainstay of the scientific realist worldview approximately a century before.

Such adjustments in what are considered to be the essential properties of the members of an ontological category represent a relaxation in the criteria for membership of that particular category, rather than the straightforward denial of such entities' existence. However, the broader and less restrictive the criteria for being such an entity or process, the more difficult it is to clarify what such entities are and the lesser the content there is to a claim that such entities exist in the mind independent world. This is, perhaps, an inevitable side effect of the loss of specificity: it is easier and less contentious to commit oneself to the existence of things than to the existence of a specific kind of thing, both in the case of fundamental ontological categories and of common-or-garden kinds (in the latter case, for example, it is more likely that animals exist than West African black rhinos). But for the realist, this loss of specificity is important and dangerous, since it renders the ontologically robust realist claims about what exists into an ever vaguer doctrine, ultimately about 'something-or-other existing', which falls far short of the thesis that our current scientific theories tell us in detail what the mind-independent world contains.

The Problem of the Inconstant Constants

The second example of potential conflict between scientific theories and scientific realism's ontological commitments concerns robust scientific realism's commitment to the existence of mind-independent natural properties, universals, tropes, types or natural kinds.[23] The stronger definition of scientific realism is explicitly committed to the existence of one variety or another of these

[23] For the purposes of this discussion, I will treat these as different names for members of the same category of entities, since they all share the functional role of providing the ontological grounding for the qualitative division of the mind-independent world and often share the criterion of being individuated in virtue of their causal roles. Whatever ontological differences there are between them are not relevant to the arguments I will make, so I will use 'property' as a neutral term to stand for the members of whichever of these categories the reader prefers.

entities, however the weaker version also presupposes that the world is objectively divided into determinate types or kinds which determine its causal structure, since these entities just are the 'essential unobservables of well-established current scientific theories', entities to which the fundamental predicates of our scientific theories refer. Put simply, the chief claim of scientific realism is that the classificatory structure of our theories for the most part matches the way in which the world is qualitatively divided independently of our theorizing, which in turn involves the claim that that the world is in fact qualitatively divided in some way in virtue of an ontological category of entities such as natural properties or natural kinds.

So how could science lead to scepticism about the nature or existence of this category of entities? This issue is made more troublesome since those who offer philosophical argument against a realist account of properties and laws still accept that we can speak in terms of them to predict and explain the happenings of the world, even if the world is not objectively divided the way we think it is (or, indeed, divided at all). Moreover, it seems difficult to imagine a scenario in which an empirical theory could have could have implications for the ontological status of properties and the laws, either probabilistic or deterministic, which relate them. In comparison with the previous discussion, this matter is quite contentious and extremely speculative, since it is based upon some novel and deeply controversial suggestions in cosmology, itself a highly theoretical area of science. However, there is perhaps no less speculation going on here than there was considered to be in the early days of quantum theory and, as I mentioned previously, it should be incumbent on the formulations of the basic categories of metaphysics to at least be compatible with what might physically be the case; that is, how the world may turn out to be on empirical investigation should not be prejudged by our ontological presuppositions. So, in the interests of speculation, I will now consider the implications of the debate in current physics about whether the so-called constants of physical theory which define the quantitative ratios between physical properties remain stable over time.

For example, there have been recent suggestions that alpha, the electromagnetic coupling constant—otherwise known as the 'fine-structure constant'—which governs interactions between charged particles and electromagnetic fields has been increasing over time. The empirical basis for this claim is that the way in which light from distant quasars is partially absorbed by dust clouds on its way to earth is different from what would be expected unless alpha has been altering slightly over the past six billion years.[24] From a more theoretical standpoint, some cosmologists have suggested that a theory which is compatible with such alterations in fundamental physical constants would provide an elegant solution to the horizon problem; that is, how distant regions of the universe are homogeneous when, given the finite speed of light, they could not be causally connected with each other.[25] Since the value of alpha is related to other important physical constants, such as the speed of light c and the charge on an electron e,[26] a change in alpha would entail a change in at least one of these constants too, which in turn would alter the quantitative ratios between other fundamental physical properties such as mass and energy. For instance, an increase in alpha might entail a decrease in the speed of light c, a constant which appears in one of the most famous identities of fundamental physical theory, $E = mc^2$, that describes the relationship between matter and energy and these properties in turn are related to many other physical properties. Clearly there are very broad implications for the relationships between the properties of physical theory should the constants be altering in some such way.

It is not important to the argument that $E = mc^2$ might turn out to be false: as long as c appears in some physical identity statement, my main point would still hold. However, those who have suggested that c alters over time do not think that this falsifies $E = mc^2$, but that the quantitative relationship between matter and energy was different in the past. On a realist construal of the theory, they are not therefore suggesting that the properties are changing what they do, rather that the 'intensity' with which properties have their customary—perhaps essential—effects has altered over time. The reason this creates difficulties for the scientific realist is that, on his

[24] See J. K. Webb, M. T. Murphy, V. V. Flambaum, V. A. Dzuba, J. D. Barrow, C. W. Churchill, J. X. Prochaska and A. M. Wolfe, 'Further Evidence for Cosmological Evolution of the Fine Structure Constant', *Physical Review Letters* 87 (2001), 091301.

[25] See H. B. Sandvik, J. D. Barrow and J. Magueijo, 'A Simple Cosmology with a Varying Fine-structure Constant', *Physical Review Letters* 88 (2002), 031302; also, J. Magueijo, *Faster than the Speed of Light: The Story of a Scientific Speculation* (Cambridge, Mass.: Perseus Publishing, 2003) for a more irreverent account of his research into VSL (Varying Speed of Light) theories.

[26] More precisely, alpha = $e^2/(\hbar \cdot c)$, where $\hbar$ = h (Planck constant)$/2\pi$. $\hbar$ might also be affected, but I am simplifying the matter a little here. Since alpha is dimensionless it is independent of any units.

account of mind-independent properties, properties are not the kind of entities which change. Moreover, the nature of a property—on the most plausible view of them—is constitutively determined by the relations in which it stands to other properties, by the role it plays in instantiating causal, and perhaps also structural, laws.[27] If the constants of the universe are not indeed constant, then the quantitative, if not the qualitative, causal relationships between the physical properties which our science postulates are unstable—by the causal definition of properties, the properties of the world are gradually changing—and this conflicts with the realist account of a world objectively divided into immutable qualities which determine the natural causal order of the world.

Our physical theory, then, works with an account of properties which permits them to be mutable, while realist metaphysics prefers to treat natural properties as immutable in order to provide constitutive accounts of objective similarity and difference between spatio-temporally distinct entities and of the mechanism of objective change. The former is important for the realist to ground accounts of what it is that spatio-temporally distinct, but qualitatively similar entities share, while the latter permits the explanation of mind-independent change without requiring the absurd claim that every time a change occurs, one thing is destroyed and another created. (This latter problem was a particularly pressing theme in presocratic philosophy which seems to have lacked a distinction between an object or a substance and the attributes it has. Thus, an object changing from hot to cold was regarded as a case of the annihilation of a hot object and the creation of a cold object, rather than of a particular persisting object losing some properties and gaining others.[28] The distinction between a substance and its attributes, of which the realist account of properties is one version, is is generally thought to originate with

[27] The nature of a particular property might not be entirely determined by its causal or structural role, but my point will affect all those philosophical views which regard what a property does as an essential feature of it.

[28] There also appears to have been confusion about those presocratics who did seem to understand change. For example, John Tzetzes, writing much later about Heraclitus, says the following: 'Old Heraclitus of Ephesus was called clever because of the obscurity of his remarks: *Cold things grow hot, the hot cools, the wet dries, the parched moistens.*' (*Notes on the Iliad,* 126H, in Jonathan Barnes (ed.), *Early Greek Philosophy* (London: Penguin Books, 1987), 115.) Now, although there are different

Aristotle.) On this (post-Aristotelian) view, particular concrete objects can persist through change because the non-essential properties which they instantiate can vary over time. But if the entities which are supposed to provide a constitutive account of change are themselves changing, then the aforementioned presocratic difficulties associated with explaining change in objects will be reignited on another level—that of properties—and so the realist ontology of properties will be robbed of some of the explanatory power which justified its existence being presupposed in the first place.[29]

There are metaphysical advantages for the realist, therefore, to continue to treat properties as immutable. However he does so at a significant cost to his scientific realism, since this move creates a gap between his metaphysical and empirical theories of the world which scientific realism was intended to bridge: there is a current scientific theory which is formulated in terms of a category of theoretical entities which either could not, or simply does not, exist in his metaphysical system. Such a category of mutable entities *could* not exist for those who, like David Lewis, treat the existence of immutable natural properties as a necessary truth—as constituents of every possible world[30]—and thus the above debate about the 'constants' of physics cannot be realistically construed within their favoured metaphysical ontology. If such philosophers wish to retain their scientific realism, then they are unable even to entertain the possibility that those physicists who suggest that constants are changing might be right.

On the other hand, those who treat the mind-independent existence of immutable properties as a contingent truth can at least find the possibility that natural properties are mutable a coherent one. But, as was the case in the debate about the correct formulation of quantum theory and whether causal laws are deterministic, the switch for the scientific realist from treating the properties referred to in our theory from being immutable to being mutable represents a considerable change in the nature of a fundamental ontological

---

metaphysical accounts of change, we would not consider the idea that an entity can persist through change by changing its properties as being obscure.

[29] See my 'Deepening the Controversy over Metaphysical Realism', *Philosophy* 77 (2002), 519–541 for further discussion about accepting metaphysical assumptions on the basis of inference to best explanation.

[30] Except, perhaps, as elements of a few extremely 'despicable' ones.

category. The question of whether we are talking about the same entities, or about a new category, could keep some philosophers engaged for years in what might ultimately be a terminological spat. But the possibility of such a debate does raise serious questions about what a realist about properties thinks he is talking about; and, for the scientific realist, about whether we have any more than the vaguest grasp upon which species of entities the predicates of our scientific theories pick out, and thus what the content of the scientific realist's claim is that they pick out mind-independent entities at all. The recognition that the characterization of such entities might not be the correct one to fit in with our current empirical theories makes scientific realism about those current theories all the harder to maintain, given that the plausibility of scepticism about the fundamental category of properties which our current theory utilizes is not merely a distant philosophical possibility but one under current empirical consideration, and thus not something that the scientific realist can choose to ignore.[31]

There are also significant metaphysical complications involved in responding to this problem by removing the condition that the natural properties, in virtue of which our empirical theories classify the world, are immutable entities. If one permits that the quantitative ratios between them do change over time, while the qualitative relations between them remain constant so that laws of nature continue to hold, one could retain a causal or nomic criterion of property identity, although the intensity or magnitude of a particular property's effects would only be determined by synchronic likeness of causal role; that is, the quantitative relations it bears to other properties at a particular time. But these revisions are unlikely to be sufficient however, since then the scientific realist would be faced with further metaphysical questions about *why* properties change in the way that they do. One could refuse to answer that question and treat it as brute fact, but that response seems methodologically, and ontologically, unsatisfactory for two reasons. First, even if it is true of the actual world that the

[31] The possibility that the characterization of properties might not be correct might bring with it the danger that many philosophical positions outside purely metaphysical enquiry might stand in need of revision since they presuppose properties: properties (under this new conception) may alter the philosophical landscape by behaving in a different way, or may require replacement by some other category since they may not be up to their former task. It is equally possible, however, that some traditional philosophical problems may be resolved, were properties to become mutable (say).

quantitative relationships between properties alter, it still seems that it might be false, and properties immutable, in possible situations sufficiently physically dissimilar to our own. For a realist, it seems a metaphysically reasonable enough question to ask in what this 'sufficient physical dissimilarity' consists, what it is about the ontological structure of the actual world which permits that, or determines how, such change takes place. Second, the realist's revised ontological picture will most probably require the postulation of such an additional underlying mechanism which determines how properties change in order to understand objective changes in ordinary objects; it will not do for the change in the world to be explained in terms of entities whose natures alter over time unless we have some understanding of how they change. But, although postulation of such a mechanism might be acceptable to the scientific realist on the basis of inference to best explanation, this would add an extra ontological feature about which our scientific theories would most likely have nothing to say.

Finally, it is worth noting here that the scientific examples which I have used against the metaphysical categories accepted by robust scientific realism are not the only ones; in fact, examples of this type are perhaps not even scarce, and, although consideration of other ones in detail would make this paper too lengthy, I hope to discuss them elsewhere.[32] In brief, a third instance of a clash between scientific theorizing and the traditional categories of metaphysics might include the difficulties of accounting for wave-particle duality in conjunction with the usual ontological distinction between particular objects and events. If one chooses to maintain objects and events as distinct ontological categories, then one runs the risk that the way in which physics classifies the world does not fit with one's metaphysics.[33] Alternatively, one can decide that the common-sense distinction which has traditionally been honoured by metaphysics does not mark a real ontological

[32] I consider some issues to do with science and the supposed unity of the world's ontological structure according to the realist in 'Disorder at the Border' (op. cit.) including Margaret Morrison's example of how the unification of electromagnetism with the weak nuclear force to produce the electro-weak theory stands against the commonly held assumption that successful reductions support the case for realism about the unity of the world's causal structure. See her 'Unified Theories and Disparate Things', *Proceedings of the Biennial Meeting of the Philosophy of Science Association.* 2 (1994), 365–373.

[33] This option seems to be acceptable to D. H. Mellor, *The Facts of Causation* (Cambridge: Cambridge University Press, 1995), 124.

distinction but that the difference between persisting entities or substances (objects) on the one hand and things which happen (events) on the other is comparative and a matter of degree, such that the latter can only be understood as existing in relation to the former and vice versa. On this ontologically interdependent account, entities which exhibit wave-particle duality are unproblematic, since the distinction between being a wave and a particle is a grammatical one, not one with any real ontological import for the realist to be troubled about, but then the tacit classification of entities as objects rather than events (or vice versa) in the rest of science is equally lacking in any objective ontological grounding.[34] Fourth, and even more briefly, one might treat Nancy Cartwright's discussion in this volume about causation as being indicative that we do not have a good unified conception about what causation is, and therefore that we cannot treat it as a metaphysically robust, objective phenomenon which exists mind-independently.[35]

4. Conclusion

The difficulties raised for the scientific realist in the previous section is that the examples show science to be more metaphysically flexible than the realist ontology with which he claims science is linked, and the realist cannot respond to the metaphysical developments of scientific theory without diluting either his realism or his commitment to the methods of science (an attitude which might once have been classified as empiricism, but what could now perhaps be called his commitment to naturalism). The conjunction of his scientific and metaphysical commitments have impaled him upon the horns of a dilemma.

On the one hand, he can emphasize naturalism and accept that there are examples such as those I have given; that is, science can show us that there is more in the world than his metaphysics allowed for. But then he is forced to maintain the rather odd

[34] This account of the ontological interdependence between particular events and objects can be found in the later work of Donald Davidson; see, for example, 'The Individuation of Events' in his *Essays on Actions and Events* (Oxford: Oxford University Press, 1985), 174–5.

[35] Once again, if one accepts Cartwright's arguments, one might conclude that the apparent plurality of causal relations means that there is no causation, that there are many types, or that we must adopt a minimal conception of causation which can embrace them all.

position that he no longer knows, perhaps even in the loosest terms, how to characterize the entities or processes he is being a realist about, since he must accept that future science could undermine these characterizations to such an extent that a serious question arises about whether we are talking about the same category of things. For example, he might assert that, in the main, the predicates of our current science pick out the properties which the world contains while not being able to say much at all (if anything) about what a property is. I contend that in such a case, the claim to scientific realism lacks content: the realist doesn't know what he's being a realist about.

Moreover, if one adopts this position of promoting naturalism over realism, it will further undermine the standard realist responses to the sceptical problems concerning 'failed' scientific predicates discussed above. Namely, 'historical' cases of reference failure in science can no longer be written off as unproblematic mistakes akin to our ordinary perceptual errors about observable entities, since these required the realist to assert that theoretical terms referring to non-existent entities such as phlogiston (say) were merely temporary errors concerning which natural kind or property was being picked out by our theory, rather than more serious errors about which category of entities our theories were about. Once it becomes questionable that the realist knows the general nature of the entities which our scientific theories are talking about, an additional 'layer' of scepticism is added that jeopardizes the realist's claim that our scientific theories tell us what is really there.

On the other hand, the realist could opt for the other horn of the dilemma and sustain realism about the entities central to our best *current* science—that is, maintain the widely accepted metaphysical realist picture and the entities with which today's realist metaphysicians are familiar—while admitting that he does not really trust scientists to do the theorizing about what there is. Admittedly, there is some motivation for adopting this strategy: some scientists are prone to making rash philosophical pronouncements, over-enthusiastically postulating entities which defy empirical discovery and making other such fast and loose ontological claims. But, although this may be reason to treat the conceptual developments of scientific theorizing with care, I do not think that it justifies ignoring them all together. However, if one does choose to do so, it seems at best confused and at worst disingenuous to call this view '*scientific* realism' at all: despite the central place it gives our current science, the methodology of scientists is not to be trusted in case it overturns the metaphysical groundwork the realist has done (and thereby, one assumes, undermines the other philosophical theories which implicitly presuppose this metaphysical account of what the objective world contains). In this case, realism about the entities in our current empirical theory will be maintained at the cost of future conceptual developments in science, which both seems to go against the spirit in which scientific realism is usually intended and, in view of past examples of metaphysical change being prompted by empirical science, to be a rather implausible position to adopt.

Despite the popular support which scientific realism enjoys, I conclude that the two terms do not sit easily together: it is not clear that one can be faithful to the spirit of both science and realism at the same time, at least when one is committed to realism in its robust form. One cannot maintain allegiance to both the methods of scientific theorizing, especially the conceptual changes it permits, and claim that our current theories tell us in any more than the vaguest terms what is really there.

5. A Final Caveat?

Perhaps a scientific realist could object that a compromise could be reached here, that one need not make a choice between science without realism, or realism without science; rather, one could have the best of both worlds by dropping the claim that scientific realism is true of our *current* science in favour of holding out for the future success of science, the formulation of a theory which has been freed by persistent empirical investigation from the kinds of difficulties which I have raised. This would remove the tension created between the inflexibility of our current realist metaphysical picture of the world and potential developments in empirical theorizing which have created problems for the scientific realist.

This suggestion raises two immediate concerns, however: firstly, whether there is such an ideal future theory to be had (even in principle); and secondly, whether the existence (in principle at least) of such a theory would sustain scientific realism, in the robust form with which we started. The first issue concerns a familiar problem about the completability of science, but not in a straightforward manner: it might not matter to the scientific realist that we could ever reach the 'end' of science, such that any event (or, any 'natural' event) could be explained at least in principle, since the scientific realist might settle for science giving an incomplete account of what there is in the world. Nevertheless, the completeness of the future theory would most probably be preferable from an epistemic perspective, otherwise it would be difficult to know whether such a 'final' theory had been reached. (An incomplete final theory would at least have to give an account of why we could not explain what it missed out, for example.) However, even armed with such a 'finished' future theory, the second difficulty concerns whether it would be unique, or unique enough, for the scientific realist to make the claim that most of the unobservables of such a theory exist mind-independently. The problem here is that if there are alternative empirically equivalent theories whose predicates are regarded as referring to mind-independent entities, then these will give different, mutually exclusive accounts of what there is in the mind-independent world. If our future theory has empirically equivalent competitors, then it does not sustain scientific realism since, robust scientific realism's commitment to metaphysical realism—the thesis that there is a single objective way in which the world is—cannot permit more than one account of what the world contains to be correct. However, if one is not prepared to accept a realist construal of our current theories, then the requirement that there is one unique-ish final theory is a difficult condition to satisfy, since it requires giving a plausible account of why it should be the case that distinct current theories will ultimately converge towards one, unified account of what the world contains, the truth (or approximate truth, if sense can be made of that notion).[36]

In this paper, I will not be able to do justice to the complex issues raised in the previous paragraph; so it will have to suffice to say that if they cannot be resolved in the robust scientific realist's favour, then even the provision of a seemingly reasonably complete future theory would not be sufficient to sustain scientific realism in its robust form. One could still call oneself a 'realist' about the theoretical entities invoked by such a theory—that is, one could take a realist attitude towards them—but it would be a mere coincidence (and an unlikely and unconfirmable one at that) that the theoretical terms within it picked out objectively existing entities or carved out the objective structure of the world. Rather than sharing the strong ontological commitments of robust scientific realism, this account of the relationship between science and what the world contains is more akin to Hilary Putnam's 'internal realism', Arthur Fine's

[36] See my 'Deepening the Controversy over Metaphysical Realism' (op. cit.) for further discussion of this issue.

Sophie R. Allen

'natural ontological attitude', or Bas van Fraassen's overt agnosticism about the objective existence of the entities seemingly picked out by our theoretical terms, and all of these views fall far short of the robust realism about the entities described by our current scientific theories which was the subject of this paper. For most robust scientific realists these views represent too great a compromise in the direction of anti-realism and do not deserve to be named 'realism' at all.

That there is a weakened, more minimal form of realism to be had which does not inevitably collapse into anti-realism, idealism, or verificationism is another important discussion, the details of which will also have to be postponed. The question here is whether one can accept the arguments of this paper that the realist does not have a good, sustainable grasp on even the metaphysical basis of his realism, the categories of entities which he presupposes to exist mind-independently, and yet still remain committed to the claim that our scientific theories are, in some sense at least, about something which exists independently of human theorizing. For example, one might claim that the types or kinds of things which the theory invokes, the systems of classification it utilizes, are at least partially determined by the theory and thus by those formulating it, while what is being classified exists independently of those doing the formulation. However, this view will most probably require some kind of defence of the scheme-content distinction on the one hand[37] and, on the other, the refutation of the commonly-held view that the rejection of the mind-independence of entities implies their complete mind-dependence and thereby entails the rejection of realism. On this latter point, I think arguments can be brought to bear against linking the entities talked about by our scientific theories too closely to the human mind which are similar in spirit to those I have already used against those views which link the terms of our scientific theories too closely to the objective world. In particular, verificationist, idealist or anti-realist conceptions of science do not permit scientific theories to be conceptually flexible enough to account for radical theoretical change, the need for theoretical reformulations which can only plausibly be accounted for by the existence of something external to the theorizers which prompts and directs them. This difficulty

[37] For one attempt to defend this distinction against the arguments of Donald Davidson and Richard Rorty, see Maria Baghramian, 'Why Conceptual Schemes?', *Proceedings of the Aristotelian Society* (1998), 287–306.

is, in essence, a close relation of the old problem of how error is possible which afflicted idealism from Berkeley onwards, and which he solved by claiming that his ideal entities are ultimately dependent upon the mind of God. For those unprepared to take this step, the problem of error might be sufficient to swing an idealist or anti-realist account of scientific ontology back in a minimally realist direction.

Miracles and Models: Why reports of the death of Structural Realism may be exaggerated

JOHN WORRALL

Introduction

What is it reasonable to believe about our most successful scientific theories such as the general theory of relativity or quantum mechanics? That they are true, or at any rate approximately true? Or only that they successfully 'save the phenomena', by being 'empirically adequate'? In earlier work[1] I explored the attractions of a view called Structural Scientific Realism (hereafter: SSR). This holds that it is reasonable to believe that our successful theories are (approximately) *structurally correct* (and also that this is the *strongest* epistemic claim about them that it is reasonable to make). In the first part of this paper I shall explain in some detail what this thesis means and outline the reasons why it seems attractive. The second section outlines a number of criticisms that have none the less been brought against SSR in the recent (and as we shall see, in some cases, not so recent) literature; and the third and final section argues that, despite the fact that these criticisms might seem initially deeply troubling (or worse), the position remains viable.

1. The Attractions of SSR

Quantum Electrodynamics predicts the magnetic moment of the electron to a level of precision better than 1 part in a billion. How, it seems natural to ask, could a theory make a prediction about what can be observed that turns out to be correct to such an amazing degree of accuracy, if what it claims is going on 'behind' the phenomena, at the level of the universe's 'deep structure', is not itself at least approximately correct? This may be logically possible, but it seems none the less monumentally implausible.

---

[1] J. Worrall, 'Structural Realism: The Best of Both Worlds', repr. in *The Philosophy of Science*, D. Papineau (ed.) (Oxford: Oxford University Press, 1996), 139–165.

John Worrall

To cite another well-worn example: Fresnel's wave theory of light—that light consists of periodic disturbances transmitted through an all-pervading elastic medium, the 'luminiferous ether'—turned out to predict, completely surprisingly even to Fresnel himself, that if a small opaque disc is held in the light diverging from a point source, then the very centre of what would be the shadow of the disc if geometrical optics were true must in fact be illuminated (indeed just as strongly illuminated at that centre point as if no opaque obstacle had been held in the light beam). This consequence of the theory was, according to an often-told story,[2] regarded by Fresnel's peers as so absurd that his theory was in danger of being laughed out of court (or at least out of the French Academy's prize-competition on the diffraction of light). But Fresnel and Arago performed the experiment with the opaque disc and lo and behold the white spot exists! How, it seems natural to ask, could Fresnel's theory correctly make a prediction that is so at odds with what 'background knowledge' would lead us to expect, *unless* it had somehow or other latched on to the way that light really is? The theory, it seems natural to conclude, must be at least approximately correct if it can get such a striking prediction right.

This is the consideration that makes most of us (and this includes the great majority of scientists) incline toward some version of scientific realism. It has often been dressed up (following Hilary Putnam) as 'the No Miracles Argument' (NMA). If Fresnel's theory, say, were substantially off-beam in what it asserts is going on 'behind' the phenomena in order to produce them, then we would, it seems, be forced to believe that the theory 'just happens' to be correct in predicting effects like that of the 'white spot'—to believe that, despite being quite false, the theory 'just happens' to have consequences about these observable situations that seem so unlikely to be correct but in fact turn out to be so. We would be forced, that is, to accept that the theory's success with this and other predictions was a mere coincidence or 'miracle'. But, so the NMA goes, we should not accept that miracles have happened, at any rate not if there is an alternative non-miraculous explanation. And in this case the assumption that Fresnel's theory

---

[2] For the real story see my 'Fresnel, Poisson and the White Spot: The Role of Successful Prediction in the Acceptance of Scientific Theories' in G. Gooding *et al.* (eds.) *The Uses of Experiment* (Cambridge: Cambridge University Press, 1989)—but the facts about the history do not affect the issues tackled here.

itself is correct or at any rate approximately correct is exactly such a non-miraculous alternative explanation of its striking predictive success. If the theory were, in particular, outright true then it would of course be no coincidence at all that what it entails about the 'white spot' is correct—all deductive consequences of a true assertion are bound to be true.[3] Hence, the NMA concludes, the reasonable assumption is that Fresnel's theory is indeed (at least approximately) correct. And the same goes for any other theory that has enjoyed comparable striking predictive successes (as all accepted theories in 'mature' science have, since this is a precondition of acceptance).

A lot can be said about the NMA (or, rather, 'the' NMA—since on more detailed analysis it splits into a number of alternative arguments). I shall say some of this later in this paper and my 2007 book[4] goes into greater detail both about its formulation and the role of 'approximate' rather than outright truth within it. However, there is no denying that the intuitions underlying the NMA are powerful and make scientific realism (in some version or other) very attractive.

The (very substantial) fly in the ointment, however, soon becomes apparent when we think some more about, for example, the 'white spot' case. Fresnel's theory, from which this startling and startlingly correct prediction was made, states that light consists of periodic vibrations transmitted through an all-pervading mechanical medium—in Fresnel's final version of the theory this 'luminiferous ether' is held to be an elastic solid. Yet Fresnel's theory was later replaced by Maxwell's electromagnetic theory of light. Maxwell's theory states that light consists of periodic changes of the electric and magnetic field strengths. In what might be called its mature form (which became definitive), this electromagnetic field is *sui generis*: it is just a basic irreducible fact about space that at each point of it and at each instant of time there are well-defined values of the electric and magnetic field strengths; the 'mature' theory explicitly denies that these field strengths can in turn be

[3]   The situation is, in fact, not so clear once it is accepted that we can (at best) claim only 'approximate' truth for even our best theories: clearly an approximately true theory is strictly speaking false and hence will have infinitely many false consequences. Here I shall avoid these complications and simply assume that if a theory is approximately true in the appropriate sense then it is no miracle that it gets some prediction correct to within observational accuracy.

[4]   J. Worrall, *Reason in 'Revolution': A Study of Theory-Change in Science* (Oxford: Oxford University Press, 2007).

John Worrall

explained via the contortions of some underlying mechanical medium.[5] Hence this later theory, it seems, straightforwardly denies the existence of the most central theoretical (alleged) 'entity' in Fresnel's theory.

This is why Fresnel's theory lies on Larry Laudan's celebrated list of theories that were predictively successful, but which we now 'know' to be radically false.[6] (Laudan plausibly argues that being inconsistent with a theory that science eventually comes to prefer is a sure sign of the falsity of the earlier theory; and that, although the notion of 'approximate truth' remains notoriously vague, if the later theory denies the existence of any 'entity' whose existence is central to the earlier theory, then no sensible account could make that earlier theory count as even approximately true in the light of the later one.) Maxwell's theory in turn was eventually replaced by a theory that makes light consist of photons—weird 'particles' lacking rest mass (and, most of the time, any definite spatial position) that obey an entirely new and probabilistic quantum mechanics. Yet both Maxwell's and the photon theory equally well entail the existence of the white spot and indeed go on to make further impressive predictions of a kind impossible to conceive within Fresnel's theory (in the case of Maxwell, about, for example, the effects of passing a beam of polarised light through an intense magnetic field).

Similarly, many people in the 18th and 19th centuries believed that Newton's theory of mechanics plus gravitation had revealed the truth about the universe (scientists were wont to lament that there was only one truth about the universe and Newton had deprived them of the opportunity to discover it)—this was in large part because of that theory's own impressive predictive successes (with, for example, the precession of the equinoxes, the 'perturbations' from Keplerian ellipses and later, of course, the discovery of Neptune). And yet Newton's theory is based on the assumptions

[5] As is well-known, Maxwell himself continued throughout his life to hold that the field must in the end be the product of an underlying material medium. However, in what might be called the mature version of Maxwell's theory, the field is indeed *sui generis*.

[6] L. Laudan, "A Confutation of Convergent Realism" repr. in *The Philosophy of Science*, D. Papineau (ed.) (Oxford: Oxford University Press, 1996), 107–138. Of course what the claim that we now 'know' those earlier theories to be false means is that our current theories (which are objectively better supported than their predecessors) imply that they are false. (Just how 'radically' false they imply them to be will be an issue that looms large in what follows.)

that space is 'flat' and infinite, that two events are either simultaneous or they are not and that there is action-at-a-distance: all assumptions that are outright denied by the theory—of general relativity—that we now accept.

It seems, then, that the facts about theory-change in science show that if it counts as a miracle for a false theory to enjoy striking predictive success, then such 'miracles' occur, if not exactly all the time, then nonetheless with some regularity in the history of science. It is surely true that if, in the light of apparently radical theory-changes like these (so-called scientific revolutions), we are forced to admit that there is no element of continuity between theories accepted at different stages in science, then the NMA is rendered impotent. The history of science in that case would display, just as Laudan argued it does, a whole series of theories that can only be counted as 'radically false' in the light of theories now accepted and yet which enjoyed unambiguous and striking predictive success of the kind pointed to in the NMA. It seems difficult to resist the suggestion that it is not just logically possible, but a possibility that we need to take seriously, that our currently accepted theories will themselves eventually be replaced by theories that stand in the same relation to them as those currently accepted theories stand to the previously accepted ones and therefore, on the supposition we are now making, will look radically false.

This is clearly not a deductively compelling inference from the historical facts concerning theory-replacement—that is why it is usually referred to as the 'Pessimistic *Induction*'. It is of course logically possible that although all previous theories were false, our current theories happen to be true. But to believe that we have good grounds to think that this possibility may be actualised is surely an act of desperation—it seems difficult indeed to supply any halfway convincing reason to hold that we can legitimately ignore the possibility that the future history of science will be similar to the past history of science and therefore to ignore the possibility that our current theories will eventually be replaced in the way that they themselves replaced their predecessors.[7] If these theory-changes

---

[7] Of course everyone believes that our theories are improving—in the sense at least that later theories are better empirically supported than their predecessors (the so-called phenomenon of 'Kuhn loss' of empirical content being a myth). But this is clearly a question of degree, while in order to justify rejecting the conclusion of the 'pessimistic induction' we would surely need some reason to think that there was a difference in *kind* between earlier theories and the present ones. As it stands, rejecting the

can indeed only count as 'radical'—that is, there is no substantial carry-over, no substantial 'continuity' from one theory to the next so that the earlier theory can only be counted as plain false in the light of its predecessor—then any form of realism seems patently untenable. Only the most heroic head-in-the-sander could then really hold that our current theories can reasonably be thought of as true. If Fresnel's theory can only count as radically false in the light of current theories of light and there is no sense in which that theory is retained (or 'quasi-retained') within those current theories, then to hold *either* that our current theories are true and will never be replaced in the future *or even* that they are approximately true and will be substantially retained within any successor theories that may come along would be a matter of pure, a-rational faith.

Various responses have been developed to the 'Pessimistic Induction'. One general line is to restrict the scope of realism to the level of theories that can be argued to have been entirely unaffected by 'scientific revolutions'. Science may now have radically different views about the fundamental constitution of matter than it did at the time when the chemical elements were thought of as consisting of billiard ball atoms equipped with a number of hooks, but science continues to tell us that one molecule of water consists of two atoms of hydrogen and one atom of oxygen. So, the suggestion goes, we should be realist about theories 'lower down' the theoretical hierarchy, but not about the fundamental theories at the top. Any such position might be called a version of 'partial realism'.

An *apparently* different, and currently widely supported, view is 'entity realism'[8]. Its proponents seem to regard this as an entirely different animal since it claims to eschew realism about theories altogether in favour of realism about entities. But how do we know (or think we know) that some (alleged) entity really is an entity—that is, how do we know (or take ourselves to know) that

---

idea that our current theories are likely to be replaced because earlier ones have on the grounds that our current theories are better supported than the earlier ones (see e.g. Peter Lipton, 'Tracking Track Records', *Proceedings of the Aristotelian Society*, Supplementary Volume LXXIV (2000), 179–205) would be rather like justifying rejecting the idea that it is likely that the current 100m sprint record will eventually be broken by pointing to the fact that the current record is better than the earlier ones.

[8]   I. Hacking, *Representing and Intervening*, (Cambridge: Cambridge University Press, 1983) and N. Cartwright, *How the Laws of Physics Lie*, (Oxford: Clarendon Press, 1983).

there is something in reality corresponding to some term involved in our theoretical framework? The answer given by entity realists is that we know this if we can *manipulate* the 'entity' in question. Hacking, for example, discussed some experiments that (*are taken to!*) involve spraying electrons at a particular kind of target and famously remarked 'If you can spray them, they are real!' It is surely patent, however, that entity realism is not a distinctive position at all but simply a (rather ill-defined) version of partial realism. One need only ask *why* we believe that we are spraying electrons at a certain target in certain circumstances or more generally 'manipulating' an electron in certain circumstances. We certainly don't ever directly apprehend the electrons, let alone the manipulation of them. The answer to this question is surely that we believe we are manipulating electrons because we accept certain *theories* that tell us that this is what we are doing and in the light of which we interpret certain observable signs (tracks in a cloud-chamber or whatever) as produced by (alleged!) electrons. Theories are inevitably involved. Entity realists are simply telling us that we should be realists about certain types of theory (ones that are sufficiently low-level and well-entrenched) and not about others (ones that are more fundamental).

Entity realism is, then, just a version of partial realism and hence it shares the main defect of that general view—namely that it surely gives up too easily on the attempt to underwrite at least some sort of realist attitude towards *our most fundamental theories*. It is these fundamental theories, after all, that are the ones that most strikingly elicit the 'no miracles' intuition. It is fundamental theories like Newton's theory of space, motion and gravitation with its prediction of the hitherto-unsuspected existence of Neptune, or Einstein's account of space-time with its prediction of the bending of the light rays by massive objects like the sun, or Fresnel's account of the basic constitution of light with its prediction of the 'white spot' or Quantum Field Theory with its prediction of the magnetic moment of the electron that provide the most striking predictive successes and, hence, the best reason that I can see for being a realist. No one should, of course, even independently of the facts about theory-change, be a fully gung-ho realist about our fundamental theories. There is, for example, a genuine current issue about whether a fully coherent version of Quantum Field Theory can even be formulated; and Quantum Mechanics and General Relativity are, to say the least, uneasy bedfellows. Hence all informed commentators expect one or, more likely, both to be 'corrected' in some not-yet-discovered 'synthesis'. No one,

John Worrall

therefore, should claim that it is reasonable to believe that our current fundamental theories are outright true (again: quite independently of the facts about theory-change); but surely one should not give up so easily on the view that it is reasonable to believe they are in some sense *approximately* true.

The only remotely plausible position that does not give up seems to me the 'Poincaré synthesis' (i.e. SSR). Henri Poincaré developed a classic account of the No Miracles Argument, but also fully recognised—long of course before Larry Laudan—the threat to any realist view that seems to be posed by the facts about theory-change in science. Poincaré wrote:

> The ephemeral nature of scientific theories takes by surprise the man of the world. Their brief period of prosperity ended, he sees them abandoned one after the other; he sees ruins piled upon ruins; he predicts that the theories in fashion today will in a short time succumb in their turn, and he concludes that they are absolutely in vain. This is what he calls the bankruptcy of science[9].

But Poincaré immediately went on to argue that this apparent threat to realism, and hence to the appeal of the No Miracles Argument, is unreal:

> [The man of the world's] scepticism is superficial; he does not take account of the object of scientific theories and the part they play, or he would understand that the ruins may still be good for something. No theory seemed established on firmer ground than Fresnel's, which attributed light to the movements of the ether. Then if Maxwell's theory is preferred today, does it mean that Fresnel's work was in vain?

> No, for Fresnel's object was not to know whether there really is an ether, if it is or is not formed of atoms, if these atoms really move in this way or that; his object was to predict optical phenomena. This Fresnel's theory enables us to do today as well as it did before Maxwell's time. The differential equations are always true, they may always be integrated by the same methods and the results of this integration still preserve their value.[10]

This might seem to amount to a ringing endorsement of instrumentalism—a view that holds that theories should be thought

9 H. Poincare, *Science and Hypothesis*, (New York: Dover, 1905), 160.
10 Op. cit. note 9, 160.

of as merely codifications of empirical results and a view that Poincaré is indeed often, but quite mistakenly, taken to hold. In fact he immediately goes on explicitly to reject that interpretation of his view:

> It cannot be said that this is reducing physical theories to practical recipes; these equations [the ones that are retained in the transition from Fresnel's theory to Maxwell's] express relations, and if the equations remain true, it is because the relations preserve their reality [more properly: we still think of them as real]. They teach us now, as they did then, that there is such and such a relation between this thing and that; only, the something which we then called motion [of the particles of the ether], we now call electric current [really: displacement current]. But these are merely names of the images which we substituted for the real objects which Nature will hide for ever from our eyes. The true relations between these real objects are the only reality we can attain ...[11].

Poincaré is claiming, in other words, that if we were to assume for the moment that Maxwell's theory of light is true, then, although we certainly could not continue to hold that Fresnel's theory was also *true*, we can continue to hold that it has correctly identified that part of the 'deep structure' of the universe that governs optical effects—because of the retention within Maxwell of Fresnel's mathematical equations governing those optical effects.

To take a straightforward example, Fresnel's theory entails that, any polarised light beam can always be regarded as the superposition of two such beams polarised in orthogonal planes and that when any light beam in air is incident on, say, a plate of glass at angle i some part of that beam will be reflected back into the air at that same angle, while the rest of it will be refracted into the glass at an angle r. His theory moreover entails the exact relative intensities of the reflected and refracted beams:

Letting I^2, R^2, X^2 be the intensities of the components polarised in the plane of reflection of the incident, reflected and refracted beams respectively and I'^2, R'^2, X'^2 the intensities of the components polarised at right angles to the plane of reflection of the incident, reflected and refracted beams respectively, then Fresnel's equations state that these variables will always be related by

---

[11] Op. cit. note 9, 161.

John Worrall

$$R/I = \tan(i\text{-}r)/\tan(i+r)$$
$$R'/I' = \sin(i\text{-}r)/\sin(i+r)$$
$$X/I = (2\sin r.\cos i)/(\sin(i+r)\cos(i\text{-}r))$$
$$X'/I' = 2\sin r.\cos i/\sin(i+r)$$

where i remember is the angle at which the light is incident on the glass (and therefore also reflected from it) while r is the angle at which the light is refracted into the glass.

These equations are retained entirely intact within Maxwell's theory. Of course, the latter theory radically 'reinterprets' the variables. In Fresnel's theory, the I, R, X, I', R' and X', which are the square roots of the intensities of the various beams, measure the maximum distance by which a particle of the elastic ether is displaced from its position of equilibrium by the passage of the wave. In Maxwell's theory (in its 'mature' form) there is no such medium and those variables instead measure forced variations in the electromagnetic field strengths. From the vantage point of Maxwell's theory, Fresnel was as wrong as he could be about *what* waves are (particles subject to elastic restoring forces and electromagnetic field strengths really do have *nothing* in common beyond the fact that they oscillate according to the same equations), but the retention of his equations (together of course with the fact that the terms of those equations continue to relate to the phenomena in the same way) shows that, from that vantage point, Fresnel's theory was none the less *structurally correct*: it is correct that optical effects depend on *something or other* that oscillates at right angles to the direction of transmission of the light, where the form of that dependence is given by the above and other equations within the theory.

The vantage point afforded by Maxwell's theory is, however, not—and almost needless to say—the *ultimate* vantage point. As Poincaré was writing, the photon theory of light was becoming generally accepted, again yielding a materially quite different view of the ultimate constitution of light than that given either by Maxwell's or by Fresnel's theory. None the less just as Fresnel's mathematical equations had been retained within Maxwell's theory, so the mathematics of Maxwell's theory was again retained (or quasi-retained, courtesy of the correspondence principle—see *below*, 142–144) in the newer photon theory. But this latest theory of light too will no doubt eventually be replaced in its turn ('pessimistic induction' about content)—though, if the history of science is any guide, the structure of that theory will be retained in later theories ('optimistic induction' about structure). As Poincaré

134

puts it, then, the various things that science, at various stages, might be thought of as telling us light *is* are 'merely names of the images which we substituted for the real objects which Nature will hide for ever from our eyes. The true relations between these real objects are the only [persisting] reality we can attain ...'.

This is why I argued in earlier work[12] that Poincaré's position—SSR—is very attractive: in a nutshell, it retains the realism suggested by the NMA (there must be *something* correct about the theoretical claims made by the theory about the 'noumena'—the theory must surely be more than 'empirically adequate') but does not assert a stronger version of realism than seems reasonable in view of the history of theory-change in science (that is, it responds adequately to the 'pessimistic induction').

2. Criticisms of SSR

SSR has, however, itself been subject to a number of criticisms in the recent literature. The main aim of the current paper is to outline (and in the next and final section respond to) just three such criticisms.

(2a) The Fresnel-Maxwell case is maximally atypical

The case that I used, following Poincaré, to motivate SSR (namely the shift considered above from Fresnel's classical elastic solid theory of light to Maxwell's theory of light as a disturbance in the electromagnetic field) is not representative of theory-shifts in the history of science in general. This has been pointed out by Colin Howson amongst others[13]—though I actually explicitly pre-conceded the point.[14] Indeed the Fresnel-Maxwell shift is so far from being representative as to be unique—or so it seems: certainly I know of, and no one else has ever cited, a 'scientific revolution' in which the mathematical equations of the earlier theory are retained *entirely intact* within the 'revolutionary' new theory, as Fresnel's equations are within Maxwell's theory. It would seem that a view of the epistemic credentials of scientific theories that claims to

[12] Op. cit. note 1.
[13] C. Howson, *Hume's Problem*, (Oxford: Oxford University Press, 2000), 39.
[14] Op. cit. note 1, 160.

John Worrall

respond to the facts about theory-change in science, but in fact responds only to one single such change, is not exactly on solid ground.

(2b) The NMA is invalid and hence the realist ingredient of structural realism is without justification

As indicated in section *1*, SSR sees the NMA (or, at least, the intuitions underlying that argument) as the main basis for being a realist about our successful theories and it then—at least apparently—qualifies that realism in the attempt to pay due regard to the facts about theory-change in science. If, then, it can be shown that the NMA can bear no weight at all—that it is a thoroughly bad argument—SSR would seem to be in obvious trouble. But Colin Howson has argued that the NMA is indeed an entirely worthless argument.[15]

Certainly if 'the' NMA were an attempt to infer deductively the truth (or approximate truth) of a theory T from its predictive success with some surprising piece of evidence *e*, it would be in obvious trouble—since it would amount in effect to a version of the fallacy of affirming the consequent. Clearly it is logically possible that a 'very' false theory could none the less happen to entail some surprising result that turns out to be correct. Indeed, if we are not too demanding about what is involved in 'predicting' a piece of evidence *e* and just take it that it is good enough if T entails *e*, then, as Howson points out, it is trivially easy to produce counterexamples to this deductive version of the argument: grueified constructions will suffice, or, in the case of mathematically formulated theories, Jeffreys-style constructions.

Suppose, to take the latter, sharper case, our theory T links two variables and is of the simple form $y = f(x)$; it predicts that when x takes the value x_0, y will take the value $f(x_0) = y_0$; while when $x = x_1$, it predicts $y = f(x_1) = y_1$; these predictions turn out to be correct when observations are made (and suppose moreover that it is somehow surprising from the point of view of 'background knowledge' that (x_0, y_0) and (x_1, y_1) are genuine data points). Can we then infer that it would be a 'miracle' if T were to get this evidence correct if it were not itself true and hence in turn that T is indeed true? Jeffreys pointed out that there are indefinitely many alternatives T'' that share this predictive success (at least in the

---

[15]   Op. cit. note 13, chapter 3.

sense that they equally well entail the data points (x_0,y_0) and (x_1,y_1)): just take T' as $y = f(x) + (x - x_0)(x - x_1)g(x)$ for *any* non-zero function $g(x)$.[16] It clearly would be strange to claim that it would be a 'miracle' if T was successful with (x_0,y_0) and (x_1,y_1) and yet was false, if, as this construction appears to show, there are infinitely many alternatives T', all of which equally entail that data and all of which equally entail the falsity of T.

Even intuitively it seems that what we would want to infer from some predictive success like the 'white spot' is not that it is *impossible* that the theory that enjoyed this success is radically false, but that it seems *extremely implausible* that it would be. A seemingly much more promising line for a formal construal of the NMA is, then, to take it to be a probabilistic argument—leading to the claim, *not* that T *is* (approximately) true, but only that it is *probably* (approximately) true.

In order to investigate this suggestion, let's first lay aside the tricky issues about approximation and operate as if our aimed-for conclusion is that some predictively successful theory T is probably true (as opposed to 'probably approximately true'). The rather nebulous talk about it being a miracle if T had got such a phenomenon as e right if it were not true seems then to translate crisply into the assertion that the probability that e would happen were T false is extremely small: $P(e/\neg T) \approx 0$. While the fact that the truth of T explains *e* can plausibly be seen as 'translating' into the claim that $P(e/T)=1$.[17] Hence the most straightforward 'translation' of the NMA into probabilistic terms seems to be:

Pr 1'. $P(e/T) =1$ (e is entailed by T).
Pr 2'. $P(e/\neg T) \approx 0$ (it would be a miracle if *e* had been the case were T not true).
Conclusion: $P(T/e) \approx 1$ and hence, given that e has occurred, $P(T) \approx 1$.

There are, of course, entirely legitimate worries about what exactly these probabilistic formulas mean; but laying these aside too for our purposes, it is easy to show, as Colin Howson again emphasises, that so long as they are indeed probabilities (that is, so long as they satisfy the formal probability calculus), then this reasoning is straightforwardly fallacious.

[16] For more details and references see op. cit. note 13, 40–44.
[17] In fact we would surely want something stronger than this probabilistic condition if we are fully to capture the *explanation* claim—not just that e is entailed by T but that T (and perhaps the 'way' in which it entails e) have some further 'nice' properties. See below 144–147.

John Worrall

Here is a simple, and by now well-known, counter example of the kind cited by Howson. Suppose that we have a diagnostic test for some disease D, and that this test (unfeasibly) has a zero rate of 'false negatives': that is, the probability of testing negative if you do have the disease is equal to 0; and moreover an (again unfeasibly) low 'false positive' rate: of 1 in a 1000, say. Suppose now that some particular person x has tested positive, what is the chance that she actually has the disease? In order to avoid changing terminology later, let T stand for the theory that a given person x has the disease, while e stands for the evidential statement that x has produced a positive result in the diagnostic test at issue. The null false negative rate is then expressed by $P(\neg e/T) = 0$; the low false positive rate by $P(e/\neg T) = 1/1000$; and the probability we are interested in—that x has D, given that she has tested positive—is $P(T/e)$.

It is often claimed to be an empirical result about human psychology that most people in these circumstances are inclined, given that there is very little chance that x will test positive if she does not have D, to infer from x's positive test result that it is highly probable that she does have the disease[18]. Such people would seem to be reasoning in perfect agreement with our latest version of the NMA:

Pr 1' holds in the diagnostic case because x is certain to test positive (e) if she has the disease (T) (i.e. $P(e/T) = 1$);
Pr 2' holds because it is extremely unlikely that x would test positive if she did not have the disease $(P(e/\neg T) = 1/1000 \approx 0)$
And the conclusion being drawn is that the probability of x having the disease in view of the positive result—that is, $P(T/e)$—is very high.

Yet, as aficionados are well aware, this inference about the diagnostic test instantiates the famous 'base rate fallacy'. *Any* non-extreme probability of T, given e, is in fact compatible with the truth of the two premises—even a probability that far from being 'very high' is arbitrarily close to zero. It all depends, of course, on the *prior probability* of T—the fallacy is to ignore this prior or 'base rate'

In the diagnostic case we can, it seems, reasonably take the prior to be the overall incidence of the disease. If the disease is very rare,

---

[18]   See for example D. Kahnemann and A. Tversky, 'Subjective Probability: A Judgement of Representativeness', *Cognitive Psychology* **3**, No. 3 (July 1972), 430–454.

a lot rarer than the rate of false positives, then the probability that x has the disease may be very low despite her positive test. So, for example, if only 1 in a million people on average have the disease, that is, $P(T) = 10^{-6}$ then the probability that x has the disease, given that she tested positive, is only 10^{-3}.

This is a straightforward consequence of Bayes's theorem; but the reason the 'posterior' is so low can, as is often pointed out, be more readily seen in an intuitive way using an urn model. Think of drawing balls at random from an urn with 1000000 balls, just one of them red (reflecting the fact that only 1 in 10^6 have disease D) and all the rest white (no disease). Each ball also has either a '+' or a '-' marked on it (corresponding to obtaining either a positive or a negative in the diagnostic test). Given that the test yields no false negatives, the unique red ball must have a '+' on it. As for the false positive rate of $1/1000$, we can't model this exactly with a integral number of balls, of course, since there are 999,999 white balls and we want a probability of one being drawn with a '+' on it to be $1/1000$, but clearly the number is close to 1000. So to a good approximation, there are 1001 balls marked '+' in the urn, all but one of which are white. So if one ball is drawn at random from the urn and it happens to have a '+' on it then there is to that same good approximation only 1 chance in 1001 that it is red. And yet something has happened, namely the patient testing positive, that we know is certain to happen if the patient has the disease and extremely unlikely to happen (only one chance in a thousand) if she does not. Pr 1' and Pr 2' both hold here, then, and yet the conclusion is (very) false. This is a clear-cut counterexample to the probabilistic version of the NMA we are considering and shows in fact that if the initial probability that some theory T is true is sufficiently low, then we can perfectly well have evidence that would be 'miraculous' were T false (probability only 1 in a thousand), and yet the probability that T is indeed false is not only not negligible but is in fact close to 1.

So far, we have taken this probabilistic argument as aiming to establish the truth of some theory T, whereas the sensible realist, already noted, wants to argue only for *approximate* truth. The prospects for producing a non-fallacious version of the NMA along these lines are, however, surely not improved by reintroducing considerations of approximate truth. As we saw, no sensible realist will want to claim anything stronger than that our current theories are approximately true, no matter how 'astounding' their predictive success. But modifying the claim in this way is not likely to help here. Let A(T) be the assertion that T is approximately true. The

John Worrall

relationship between A(T) and e is altogether less clear-cut than that between T and e. We can take it that, the relevant auxiliaries being presumed as given, T logically entails e, while whether or not A(T) entails e is unclear. Nonetheless, since the aim of the NMA is for the success with e to have a major impact on the intuitive credibility of T—here reflected in an increase in its probability—presumably its proponents will need to claim that $P(e/A(T)) \approx 1$. Again, the NMA relies on the idea that the evidence at issue would be very improbable were T not even approximately true, so realists developing this form of the argument would presumably be committed to the premise $P(e/\neg A(T)) \approx 0$. Hence we have a simple modification of the probabilistic argument

Pr 1" $P(e/A(T)) \approx 1$
Pr 2" $P(e/\neg A(T)) \approx 0$
Conclusion: $P(A(T)/e) \approx 1$ and hence, given that e has occurred, $P(A(T)) \approx 1$.

But then clearly the base rate problem kicks in just as before: depending on the value of the prior probability that T is approximately true, any posterior for T's approximate truth—including a posterior as close to zero as you like—is compatible with the truth of premises Pr 1" and 2".

It seems, then, that 'the' NMA is in trouble and hence, since the realist element of SSR is based squarely on it, so is SSR.

2(c) The 'Newman argument' 'destroys' SSR

These two arguments seem bad enough news for SSR, but the argument that seems to have convinced most philosophers of science of the untenability of SSR is still a third one. This goes back to a paper of 1928 by the Cambridge mathematician Max Newman, responding to Bertrand Russell's version of structural realism.[19] Newman's argument was brought back to the attention of philosophers of science via a 1985 article by Demopoulos and Friedman[20]. The argument in its crispest form goes as follows:

[19]   M.H.A. Newman, 'Mr. Russell's Causal Theory of Perception', *Mind* **37**, No. 146 (April 1928), 137–148.
[20]   W. Demopoulos and M. Friedman, 'Critical Notice: Bertrand Russell's *The Analysis of Matter*: Its Historical Context and Contemporary Interest', *Philosophy of Science* **52**, No. 4 (December 1985), 621–639.

1. SSR is committed to the view that the Ramsey sentence of any scientific theory T captures the full 'cognitive content' of that theory.

2. However, as Newman showed, the Ramsey sentence of any theory imposes only a very weak constraint on the universe—it amounts in essence to a mere cardinality constraint, and so if there are sufficiently many objects in the universe then the Ramsey-version of T, for any T, will be true.

3. However it is clear that standard scientific theories impose much more stringent constraints on the universe if they are to be true than merely a constraint on the minimum number of entities the world must include.

4. Hence SSR is committed to an account of the cognitive content of scientific theories that is plainly untenable and is, therefore, itself untenable.

No wonder then that the Routledge Encyclopaedia article on Russell refers to Newman's argument as the 'definitive refutation' of his structural realism:

> Newman's argument is the definitive refutation of the Structural Realism of Russell (1927) ... Russell quickly abandoned SR when Newman showed that any set with the right cardinality could be arranged so as to have the same structure as the world—a result analogous to that claimed in Putnam's model-theoretic argument against realist theories of reference (Demopoulos and Friedman 1989).

Of course this leaves it open that there is something especially faulty with Russell's version of SSR, but in all respects relevant to the current discussion this is not true (or at any rate I do not believe it to be true). If so, then advocates of SSR such as myself and Elie Zahar[21] seem to have shown reprehensible ignorance of the literature in advocating a position that had already been conclusively demolished.

3. Responses to Criticism: why reports of the death of SSR may be exaggerated

Is SSR in straits as dire as the above three criticisms seem to suggest? I consider the criticisms in turn.

---

[21] E. Zahar *Poincare's Philosophy: From Conventionalism to Phenomenology*, (Chicago: Open Court, 2001).

John Worrall

3(a) The atypical nature of the Fresnel-Maxwell shift

There is no denying that the theory-shift in optics from Fresnel to Maxwell is unrepresentative. Indeed, as indicated, I already emphasised this point in developing my defence of SSR. In all other cases, the best that can be argued is that, once a science has reached maturity,[22] the mathematics of any theory replaced in a 'scientific revolution', while not being retained fully intact, is instead 'quasi-retained' *modulo* the 'correspondence principle'.

The most straightforward cases of the application of this principle are where the equations of the older theory reappear as limiting cases of the equations of the newer theory (and moreover the limiting cases characterise the area in which the older theory had proved entirely empirically successful). A classic case is of course represented by the relationship between the Special Theory of Relativity and Newtonian physics—the Newtonian equations being recovered from the Einsteinian ones as v/c tends to 0 (where v/c tends to 0 as a body's velocity is ever smaller compared to the velocity of light; and where Newtonian and Einsteinian predictions—though always strictly different—are entirely empirically indistinguishable for relatively slowly moving objects).

As Michael Redhead has pointed out[23], not at all applications of the correspondence principle fit this pattern (one example he cites is the transition from geometrical to wave optics, though I am unsure whether geometrical optics can count as a 'mature', that is, in my terms, genuinely predictive theory). Others may well feel that the 'continuity' afforded by the correspondence principle *in general* is hardly worthy of the name. And hence that any 'realism' founded on it is, in turn, hardly worthy of the name.

[22] Larry Laudan complains (op. cit, note 6) that the notion of 'maturity' is introduced by realists as an ad hoc device: whenever it seems like there is no sense in which an earlier theory continues to look 'approximately true' in the light of its successor, the realist can claim that that earlier theory was accepted only when the science that it contributes to was 'immature'. However, as I have explained before (op cit. note 1), it seems that the realist should be ready to 'read off' her notion of maturity from the NMA which is her main support—taking it that a scientific field attains maturity once its accepted theory enjoys genuine predictive success (that is, it predicts some general phenomenon that was either unknown at the time or was not used in the development of the theory concerned).

[23] See, e.g., his 'The Unseen World' in C. Cheyne and J. Worrall (eds.) *Rationality and Reality: Conversations with Alan Musgrave*, (Springer, 2007).

It is surely true that the strength and character of scientific realism, of what the reasonable attitude is toward the purely theoretical claims of current theories, depends on how strong a notion of continuity can be extracted from the history of theory-change in science. And that's the way round it clearly has to be: science, and the history of theory-change within in, strongly constrains the reasonable philosophical view. If the notion of structural continuity via the correspondence principle is not strong enough for your tastes then you will not be happy with calling SSR a 'true' version of realism. But this seems to be a merely semantic issue. The extra ingredient that SSR adds to a van Fraassen-style empirical adequacy view may not be very strong but it *is* an extra ingredient and it comes at no real price. According to the account of theory change that underpins SSR, successive theories in science have not only been successively more empirically adequate, but there has always been a *reason*, when viewed from the vantage point of the later theory, *why* the earlier theory achieved the degree of empirical adequacy that it did—namely that the earlier theory continues to look approximately structurally correct: its mathematical equations are retained *modulo* the correspondence principle.

Given this underpinning, then SSR is just a simple and surely innocuous inductive step away: it seems reasonable to believe that currently accepted theories, if they are replaced at all (as seems highly likely), will be replaced by theories in the light of which they will continue to look approximately structurally correct. If, as seems right, we count as a version of scientific realism any view that asserts that it is reasonable to hold that our successful theories are more than simply highly empirically 'adequate'—they are empirically adequate because they (can reasonably be taken to) 'latch on' in some way to the 'deep structure' of the universe, then SSR counts as realism. It goes on to insist that the way that our theories thus 'latch on' to the 'deep structure' of the universe cannot be further specified—to suppose that it can would be to suppose that we can somehow have access to the universe that is not theory-mediated and thus can directly compare what our theories say with reality. But once articulated this supposition is clearly untenable. This will be disappointing for some. But, as I shall argue in the sub-section *3(c)*, no account of how that relationship might be further specified makes any real sense. If we are talking about coherent, defensible positions, then SSR is as realist as it gets.

In sum, to count as a fully fledged version of realism any such view must say something of a realist kind about fundamental, frontier theories, but any version of realism about fundamental, frontier theories is dependent for its plausibility on the production of an account of "continuity through revolution" in the history of science. No one should claim a stronger sense of continuity, and hence a stronger version of realism, than is compatible with the historical record. We should look for the strongest such version and see if it is a continuity worth having. If there is no such notion of continuity worth having, then there is no sustainable version of realism. However, I hold that there is a continuity (admittedly of an approximate kind) at the structural level that is substantial enough to count and hence I hold that SSR is a sustainable version of scientific realism, and indeed, as I shall try to show again later, the only sustainable version.

3(b) What should we expect from 'the' NMA?

Representations of 'the' NMA as either an attempted deduction or as an unadorned probabilistic argument are, as we saw in section *2(b)*, undeniably and straightforwardly invalid. The pro-realist intuitions elicited by particular cases of predictive success remain strong however. This suggests that there may be some other way of developing the argument that is *not* fallacious.

One standard way of running the argument (unacknowledged by Howson) is as an instance of 'inference to the best explanation'. The claim is that the (approximate) truth of the theory T is the 'best explanation' of T's success in predicting e (where, moreover, there is an implicit assumption that T does more than simply entail e, it also 'explains' it). This means that gruesome or Jeffreys-style constructions are doubly barred from counting as counterexamples. First of all, someone construing the argument in this way would allege that, although Jeffreys-style constructions clearly deductively entail the evidence at issue (in our example: the data points (x_0, y_0) and (x_1, y_1)), they do not *explain* those data points. And secondly, and of course relatedly, while it *might* be argued that the 'best explanation' of the original theory T's success with e is that it is (approximately) true (this will depend on the details of the particular case), the best explanation of the Jeffreys-style alternatives 'success' with e is that they were constructed exactly so as to entail it. Hence, although there are undoubtedly infinitely many rivals to T (in the precise sense that they entail that T is false) that equally well entail e, this does not mean that there are infinitely many rivals that equally well 'explain' e and hence that equally well generate the 'no miracles intuition' (nor does it even mean that there is *one* rival that does so).

Although this suggestion seems to me on the right lines, talk of 'inference to the best explanation' gives it an air of precision and formality that the argument scarcely deserves. The fact is that no one has much, if any, idea about how to articulate the extra requirements on a theory, beyond actually entailing some evidence e, for it to *explain* e. A high degree of 'unity' is often cited here, but, although I believe that this is indeed the crucial notion, and although it is easy to point to theories that clearly possess such unity and at others that clearly do not, no one has succeeded in giving a general characterisation of the notion. It would appear easier to characterise what a *dis*unified theory is: essentially one that has been produced from an earlier theory via *ad hoc* modifications designed to remedy defects in that earlier version (standardly empirical anomalies). But this in effect presupposes that the initial theory was unified and again while it is easy to exemplify the idea, it has not proved possible to articulate it in general terms.

What this attempted formulation indicates, however, is that there is indeed an extra implicit assumption in particular applications of the NMA beyond the fact that the theory concerned gets some piece of evidence correct. It would be a 'miracle' if a theory that got some striking piece of evidence correct *and was a theory of the 'right sort'* was nonetheless radically off-beam. Again Poincaré was fully aware of the situation:

> We have verified a simple [sic] law in a considerable number of particular cases. We refuse to admit that this coincidence, so often repeated, is a result of mere chance and we conclude that the law must be true in the general case. Kepler remarks that the positions of the planets observed by Tycho are all on the same ellipse. Not for one moment does he think that, by a singular freak of chance, Tycho had never looked at the heavens except at the very moment when the path of the planet happened to cut that ellipse ... [I]f a simple [sic] law has been observed in several particular cases, we may legitimately suppose that it will be true in analogous cases. To refuse to admit this would be to attribute an inadmissible role to chance.[24]

So the success of Kepler's law in entailing the phenomena is only one feature of the situation for Poincaré: it is also important that

[24] Op. cit. note 9, 149–150.

John Worrall

the law is *simple*. After all, Tycho's data (as Poincaré was surely implicitly aware) fits not only the Keplerian ellipse but also indefinitely many Jeffreys-style versions thereof.

One significant feature to notice is, then, that Poincaré is pointing out that something like the NMA is needed to underwrite what are generally taken to be straightforward inductions to conclusions that are often thought of as 'empirical generalisations'.

As for the NMA in general, he is acknowledging that there is an implicit, and perhaps vague, assumption about simplicity or prior plausibility. It is unlikely that something as simple as the ellipse would fit in all the many cases we have observed (and fail to fit in no observed cases) without being generally true. (It is this that makes van Fraassen's position in *The Scientific Image* so difficult to empathise with. By suggesting that we do indeed accept our best current theories as 'empirically adequate'—where this quite explicitly means empirically adequate *across the board* and not just with regard to empirical results that have already been checked observationally—he is implicitly appealing to the NMA. But why then reject the idea that that same argument has force when it concerns 'theories'?[25])

Poincaré's appeal to simplicity could, of course, be captured within a Bayesian account. And indeed Howson points out (op. cit. note 13) that, although the probabilistic version of the NMA is invalid as it stands, it is easy to make it (probabilistically) valid if, far from ignoring the base rate, one asserts, as an extra premise over and above the claim about the impact on the theory T of evidence e, that the prior probability of T is at least reasonably high.

It is not that this Bayesian rendition seems to me incorrect in any way, but rather (and as usual) it seems to supply no extra 'explanatory force'. The assumption that, in cases where the NMA-intuitions kick in most strongly, the 'prior' of the theory concerned is at least reasonably high seems just a—not especially perspicuous—way of reflecting our intuitive judgment about the unity or simplicity of that theory; and not to add anything to this intuitive judgment. (Indeed the intuitions here seem stronger: they characterise, in a way that (personalist) Bayesianism declares impossible, those cases in which it is, and those in which it is not, *reasonable* to assume a reasonably high prior. It is not a subjective

---

[25]   See B. van Fraassen, *The Scientific Image*, (Oxford: Clarendon Press, 1980).

matter that a theory claiming that planets move in ellipses is simpler than one that claims that their orbits are some Jeffreys-style monstrosity.)

All in all, one should not, it seems to me, expect too much from 'the' NMA. Neither of the ways of running it *apparently* more formally while avoiding patent invalidity—as an 'inference to the best explanation' or as a Bayesian inference with a substantial prior as extra implicit assumption—adds anything of real substance to the underlying intuition. The 'argument' ('consideration' might be more honest) should, so it seems to me, be thought of as doing little more than setting the default position: given the, occasionally staggering, predictive success of our current (and some of our earlier) theories (and given that they have achieved this predictive success without any necessity for *ad hoc* modifications), the reasonable default assumption is that they have latched on in some way to the 'deep structure' of the universe.

There is no question of logical compulsion here—just the suggestion that it seems more reasonable than not that theories that enjoy this sort of success must be 'on the right lines'. Moreover, like all default positions, this one is obviously defeasible; and it would indeed be straightforwardly defeated, if Laudan were right that there is no real continuity (even of a somewhat attenuated kind) between successive theories in 'mature' (i.e. predictively successful) science. The pessimistic induction, if accepted, would trump the NMA. This is because the history of science would then, as already noted, provide a list of alleged 'miracles'—theories that enjoyed striking predictive success but which are not even approximately true by the lights of current science. And, when it comes to 'miracles', familiarity surely does breed contempt. This is why it is important for a defensible realism to establish a way in which successive theories in mature science have indeed been at least quasi-accumulative. And I claim that only the structuralist can successfully establish such an account.

3(c) The 'Newman argument': is SSR really realism?

The Newman argument is in effect that no defensible view of scientific theories can be based on the claim that the full cognitive content of such a theory is captured by its Ramsey sentence— because that sentence imposes a mere cardinality constraint on the universe. There is no doubt that SSR *is* committed to the claim that a theory's full cognitive content is captured by its Ramsey sentence.

John Worrall

The issue then clearly is whether Newman indeed establishes that the Ramsey sentence of a theory is as weak as he seems to claim.

The Ramsey sentence of a theory T, as is well-known, is constructed by replacing all the *theoretical* predicates in T by second-order variables and then existentially quantifying over those variables.[26] The claim that this sentence captures the full cognitive content of T simply reflects the fact that, so far as our fundamental theoretical notions are concerned, we know about them only by description—that is, via their role in our theories. If asked what an electron, say, is (or rather what we *think* an electron is), one can do no more than recite our current (full!) theory of electrons— electrons (if they exist at all) are *whatever it is* that satisfy our current relevant theories. But this means that a theoretical term like 'electron' in effect plays the role of a (second-order) 'ambiguous name'; and, as is well-known, in systems of logic that employ ambiguous names α_i, $P(\alpha_i)$ is logically equivalent to there is an x, $P(x)$.[27]

In order to deny that the Ramsey sentence of T captured the full cognitive content of T, one would have to assert that we have some independent way of describing how our theoretical terms 'latch on to' or 'hook up with' reality—independent, that is, of our theories themselves. This can only make sense if something like the causal theory of reference is adopted. But whatever its attractions as an account of how we practically manage to communicate via commonly held assumptions about the reference of ordinary names, the causal theory of reference as an account of the reference of primitive theoretical terms is surely patently hopeless. Think about how one might 'ostend' the electromagnetic field, say, in order to 'baptise' it: clearly we could only know that we are ostending the field (in fact you can point in any direction you like!) via the *theory* of the field. It is just a fantasy (given credence by unthinking reflection on orthodox logical semantics) that we can "stand outside" of our theories and directly compare terms in them with a reality that we can access directly without any theory. No one *really* believes this, though many act (and even sometimes write) as if they do. But once this is recognised as the fantasy it is, then there

---

[26] Of course if you presuppose set theory then the theoretical predicates can be replaced by predicates varying over sets and then the Ramsey sentence is entirely first-order.

[27] See, for example, P. Suppes, *Introduction to Logic*, (New York: Van Nostrand, 1957).

just is no question but that the Ramsey sentence of T (in its original, rather than Lewisian form[28]) captures the full cognitive content of T.

It would then be rather surprising if Newman's argument really did establish—as so many commentators now seem to believe it does—that the Ramsey sentence of any scientific theory imposes only a cardinality constraint on the universe. (That is, that all that the Ramsey sentence of any theory requires of the universe in order for that sentence to be true is that the universe includes sufficiently many individuals.) Given that the Ramsey sentence just *has to* capture the full cognitive content of a theory, this would mean that all human theorising is doing no more than imposing a constraint on how many individuals there are in the universe and that is absurd, if anything is.

Yet this is exactly what Demopoulos and Friedman, quoting Newman, seem to assert:

> The difficulty is with the claim that *only* structure [as revealed by the Ramsey sentence] is known. On this view "the world consists of objects, forming an aggregate whose structure with regard to a certain relation R is known, say [it has] structure W; but of R nothing is known ... but its existence; ... [A]ll we can say is *There is* a relation R such that the structure of the external world with reference to R is W." (Newman 1928, p. 144). But "*any* collection of things can be organized so as to have the structure W, provided there are the right number of them" (p. 144 italics added). Thus, on this view, only cardinality questions are open to discovery! Every other claim about the world that can be known at all can be known a priori as a logical consequence of the existence of a set of α- many objects. For, any given set A of cardinality α, can with a minimum of set theory or second-order logic establish the existence of a relation having structure W, provided that W is compatible with the cardinality constraint that $/A/ = \alpha$. (The relevant theorem from set theory or second-order logic is the proposition that every set A determines a full structure, that is, one which contains every relation [in extension] of every arity on A;...)[29].

---

[28] D. Lewis, 'How to Define Theoretical Terms', *Journal of Philosophy* **67**, No. 13 (July 1970), 427–446. Lewis's argument for reading the quantifiers as 'there is a *unique* Ø such that' seem to me entirely unconvincing.

[29] Op. cit. note 20, 627–628.

John Worrall

The most obvious mistake that Newman makes (and he seems to be followed, at any rate in this passage, by Demopoulos and Friedman) is to assume that in order to identify the structural claims of a theory T one should replace *all* the predicates occurring in it by predicate variables and existentially quantify over them—that is, that all the predicates in the theory should be treated as theoretical.[30] It is surprising that Russell should have conceded to Newman (as Demopoulos and Friedman correctly report that he did), given that his whole structural realist view was based on a sharp distinction between theoretical and observational notions (in his own terms between things known *by acquaintance* and things known only *by description*).

It is true that if *all* predicates are quantified-over, then the resulting Ramsey sentence is hopelessly weak (though the situation, as we'll see, is rather more complicated than Demopoulos and Friedman, following Newman, assert). But once it is recognised that some predicates are to be taken as observational and therefore as interpreted independently of theory and so emphatically are *not* to be quantified-over, then the charge that the Ramsey sentence of any theory imposes a mere cardinality constraint on the universe is easily refuted. It is in fact well-known that, in that case, the original theory T and its Ramsey sentence will be empirically equally powerful: that is, every sentence that is expressible purely in observational terms and is deducible from T is also deducible from its Ramsey sentence. Thus the Ramsey sentence of T is co-refutable with T; clearly then the Ramsey sentence of any theory that has empirical content imposes much stricter constraints on the universe than any mere cardinality constraint.

(If we suppose, as Newman himself seems to have done, that there is no distinction to be made between observational and theoretical predicates in T, then T's Ramsey sentence will indeed fail to capture its full cognitive content—being in effect a purely mathematical statement. However even then Newman's claim requires modification. He suggested that if structuralism were correct, that is the Ramsey sentence was all that could be known, then this would entail that all that can be known 'theoretically' is

---

[30] This section of the paper follows the treatment in J. Worrall and E. Zahar, 'Appendix IV: Ramseyfication and Structural Realism' in E. Zahar *Poincare's Philosophy: From Conventionalism to Phenomenology*, (Chicago: Open Court, 2001), 236–251. I am in general greatly indebted to Elie Zahar for many long discussions of structural realism and for invaluable help with some technical issues.

"the number of constituting objects"[31]. But in fact only in exceptional cases does the Ramsey sentence of a theory (formed by quantifying over *all* the theory's theoretical predicates) determine the cardinality of domains that satisfy that theory. For example if T is a contradiction then so is its Ramsey sentence; while if T is $\forall x(S(x) \leftrightarrow \neg(x=x))$ then the Ramsey sentence $\exists \Phi$ $(\forall x(\ \Phi(x) \leftrightarrow \neg(x=x)))$ is a logical truth, since $\Phi(x)$ can be chosen to the predicate $\neg(x=x)$. In the first case there are no models and in the second case all interpretations are models. In neither case is there of course any restriction on cardinality. Moreover, by the upwards and downwards Löwenheim-Skølem theorems, if T is any theory with an infinite model of any cardinality then it has models of *all* infinite cardinalities. Hence a logically true T's Ramsey sentence will be true in virtue of a range of structures with domains of all infinite cardinalities. It is only in the case of a theory that has only finite models that its Ramsey sentence determines the size of the domain of individuals. So, for example, if our theory were $\forall x\ Sx\ \&$ $\exists x_1, x_2, \dots x_n\ (\neg(x_1 = x_2))\&\dots\&\ \neg(x_{n-1}= x_n)\ \&\forall x(Sx \leftrightarrow (x = x_1)$ v $\dots$v$(x = x_n))$ then its Ramsey sentence would be equivalent to the assertion that the domain of individuals consists of exactly n members.

It should be admitted that while Demopoulos and Friedman go along with Newman's untenable claim in the passage just cited, they do elsewhere state:

> ... if our theory is consistent and *if all its purely observational consequences are true*, then the truth of the Ramsey-sentence follows as a theorem of set-theory or second-order logic, provided our initial domain has the right cardinality ...[32].

And perhaps this is their real position. It is, of course, not open to the above objection. The claim now seems to be that although the Ramsey sentence of any theory T says everything that T does about observational matters, it is somehow not strong enough to capture the full cognitive content of T.

It is difficult to know what to make of this claim. Notice that the empirical equivalence of T and its Ramsey sentence extends far beyond what we might normally take that phrase to mean. The Copernican and Ptolemaic theories, for example, are (or can be made) empirically equivalent with respect to observed planetary motions (courtesy of delicate adjustment of epicycles); but any

---

[31] Op. cit. note 19, 144.
[32] Op. cit. note 20, 635, emphasis added.

theory T and its Ramsey sentence are observationally equivalent in the much stronger sense that *any sentence* that is expressed in the observational vocabulary and is entailed by T is also entailed by its Ramsey sentence. Many sentences expressed in purely observational vocabulary should count, in any one's book, as theoretical: one often cited example is the claim that there are unobservables— that is, individuals possessing no observable property.

Moreover, the Ramsey sentence of any theory is itself observational in this sense—trivially so since all the theoretical predicates in the original theory have been removed and quantified-over. This makes the claim just quoted from Demopoulos and Friedman in one sense trivially true since of course (a) T's Ramsey sentence is provably a logical consequence of T and (b) T's Ramsey sentence, like any other sentence, entails itself. (On the other hand, if what Demopoulos and Friedman mean by 'observational consequences' are singular, decidable sentences, then what they say can straightforwardly be proved to be false using the compactness theorem.[33])

Quantifying over the erstwhile theoretical predicates removes them linguistically, but the theoretical terms surely live on within the Ramsey sentence via the structure that they impose on the observational content. (That is, after all, what structural realism is about.) Indeed, as Elie Zahar pointed out to me, if we follow Quine's dictum that 'to be is to be quantified over' (or, better: 'to be asserted to be is to be quantified over in a sentence that you assert') then the Ramsey sentence of some theory T $(S_1,... S_n, O_1,... O_r)$, where the S_i are theoretical predicates and the O_j are observational, namely $\exists \Phi_1...\exists \Phi_n$ $(T(\Phi_1,...,\Phi_n O_1,... O_r)$ clearly asserts that the 'natural kinds' S_1, ..., S_n (the extensions of the theoretical predicates $S_1, ... S_n$ in the initial theory T) exist in reality just as realists want to say. It is just that—as always—we fool ourselves if we think that we have any independent grip on what the S_1, ..., S_n are aside from whatever it is that satisfy the Ramsey sentence (assuming that the theory whose Ramsey sentence we are considering is true).

In the end, then, the much-vaunted 'Newman objection' is no objection at all to SSR. In its original form it relies on a false assumption (that SSR regards all predicates as theoretical); and in its modified Demopoulos and Friedman form it simply highlights a consequence of SSR that is essential to it. The fact that it has been found so troubling depends, I think, on nothing more than a

[33]   Op.cit. note 30, 235.

residual vague feeling that if this is what SSR amounts to then it doesn't count as *real* realism. So I shall end by confronting that question directly: is SSR 'really' realism?

The answer is 'no' if we take Putnam's characterisation. This, as is well known, has the realist assert that 1) Theoretical terms in mature science refer; and 2) what our currently accepted theories in mature science say using those terms is at any rate approximately true.[34] That is, a realist says that there definitely are electrons (for example) but then is allowed to concede that what current theories say about them may not be exactly true (though will nonetheless be close). This is how 'referential continuity' is supposed to be restored across theory-change—earlier theorists were talking about the same entity when they talked about 'electrons', even though what they said about them was different from (though in some sense nonetheless approximates to) what our current theories say. But why should a realist not be equally as fallibilist and tentative about the mode of reference of the terms in theories as she is about those theories' truth? In any event, no one seriously holds, as I have remarked several times, that we have any theory-independent access to the furniture of the universe that would allow us to compare (even 'in principle') the notions conjectured by our theories with what there really is.

Scientific Realism of any stripe consists of a metaphysical thesis conjoined with an epistemological one. The metaphysical thesis is often taken as the claim that there exists a reality independent of the human mind. But, as Elie Zahar points out, a more accurate rendition in our more enlightened, post-Cartesian-dualism times would be:

> There exists a structured reality of which the mind is a part; and, far from imposing their own order on things, our mental operations are simply governed by the fixed laws which describe the workings of Nature.[35]

I take this metaphysical thesis as a given. Scientific Realism adds the epistemological thesis:

> Not only is this structured reality partially accessible to human discovery, it is reasonable to believe that the successful theories in mature science—the unified theories that explain the phenomena without *ad hoc* assumptions—have indeed latched

[34] H. Putnam, 'What is Mathematical Truth' in *Mathematics, Matter and Method* (Cambridge: Cambridge University Press, 1975), 69–70.

[35] Op. cit. note 21, 86.

on, in some no doubt partial and approximate way, to that structured reality, that they are, if you like, *approximately true*.

If it is assumed that to be a 'real realist' one must assert that the terms in our current theories refer as part of an acceptance of a correspondence or semantic view of truth as the account of what it means for our theories to have latched on to the real structure of the world, and it is assumed that the realist must develop some sort of weakened version of correspondence as her account of 'approximate correspondence with reality' then SR does not count as 'real realism'.

But there is no reason why the way in which a theory mirrors reality should be the usual term-by-term mapping described by traditional semantics. Indeed, as I have remarked several times already, if we are talking about an epistemically accessible notion then it cannot be! SSR in fact takes it that the mathematical structure of a theory may globally reflect reality without each of its components necessarily referring to a separate item of that reality. And it takes it that the indication that the theory does reflect reality is exactly the sort of predictive success from unified theories that motivates the No Miracles Argument.[36]

SSR may well be more modest than many who have sought to defend some version or other of scientific realism might like. But the modesty involved in SSR is far from undue. No stronger version of scientific realism is either compatible with the facts about theory-change in science or compatible with any truly defensible epistemological view of how our best theories are likely to 'link up with reality'. If SSR isn't realism then nothing defensible is. But SSR is, I suggest, in fact a modest but defensible version of scientific realism—reports of its death have been greatly exaggerated!

[36] My account in this section of the paper is particularly indebted to a number of conversations with Elie Zahar.

Counter Thought Experiments

JAMES ROBERT BROWN

Introduction

Let's begin with an old example. In *De Rerum Naturua,* Lucretius presented a thought experiment to show that space is infinite. We imagine ourselves near the alleged edge of space; we throw a spear; we see it either sail through the 'edge' or we see it bounce back. In the former case the 'edge' isn't the edge, after all. In the latter case, there must be something beyond the 'edge' that repelled the spear. Either way, the 'edge' isn't really an edge of space, after all. So space is infinite.

This example is typical of thought experiments in general. We set things up in the imagination, we let it run, we see what happens, and we draw a conclusion. It's also quite similar to a real experiment, except that it's done in the imagination rather than in the real world. And like real experiments, thought experiments are fallible. In this case we would now make a distinction between *unbounded* and *infinite,* so that the conclusion Lucretius drew, we now clearly recognize, does not follow from what went before.

Lucretius is but one of many different types of thought experiments. Positive or constructive thought experiments support some theory, while negative or destructive ones undermine. My interest is in a special class of negative thought experiments that I shall call 'Counter Thought Experiments'. I'll largely ignore other types, except for the sake of contrast.

Examples of Negative Thought Experiments

One type of destructive thought experiment shows some existing theory to be self-contradictory. Einstein chased a light beam with a view to see what the wave front looked like. If we were to run on a pier toward the shore at the same speed as an incoming water wave, we would see a static hump in the water. Perhaps we would have a similar experience in the case of light, since light, according to Maxwell's theory, is an electromaganetic wave. The light wave, however, is dependent on change: a changing magnetic field gives rise to an electric field, and a changing electric field gives rise to a magnetic field. When Einstein catches up to the front of the light wave, he would see static fields, but then a light wave cannot exist.

Galileo reduced Aristotle's theory of motion to an absurdity in a rather simple but ingenious way. The first part of his wonderful thought experiment on falling bodies is a typical *reductio ad absurdum*. Aristotle claimed that heavier bodies fall faster than light ones (H > L). Suppose we attach a heavy body and a light one together. Then the combined object must fall faster than the heavy one alone (H+L > L). But the light component of the combined body will act as a drag, slowing the whole thing down so that it is actually slower than the heavy body falling alone (H+L < L). Thus, we have an absurdity, and Aristotle's theory is destroyed. Galileo's thought experiment then takes a second step, this time a positive one. It becomes obvious how bodies must fall, given the way the absurdity was achieved. All bodies must at the same rate (H = L = H+L).

Showing an internal contradiction is not the only way to undermine an existing theory. Some thought experiments show the theory to be contrary to other established (including common sense) beliefs. Schrödinger's cat is a prime example. Schrödinger took the wierdness of the Copenhagen interpretation of quantum mechanics at the micro-level and brought it to the macro-world of everyday objects such as a cat. It was bad enough that an atom could be in a superposition of two different states (e.g., energetic and decayed), but the consequence of that view seemed to imply that even a cat could be in a superposition of living and dead. This is not a contradiction. Some physicists (Wigner) actually were willing to accept it. But it is a gross violation of common sense.

In argument terms, these thought experiments show the premisses false. That is, they show that the theory in question must be false. In the first case just mentioned, the thought experiments show that there is something wrong with the conjunction of electrodynamics and basic assumptions about moving reference frames. In the second, Aristotle's view that bodies fall at rates related to their weights is wrong. In the third, the target is the Copenhagen interpretation that allows physical systems to be in reality in states of superposition.

Of course, these are debatable outcomes. One could try to save the initial theory by putting the blame on something else. Maybe there's a difference between genuine bodies and the composite bodies of Galileo, with the true laws applying only to the genuine ones. Maybe there is a micro-macro distinction with atoms going into states of superposition, but not cats. As with any real experiment, there is lots of room for rival interpretations, not to mention outright mistakes.

A second type of negative thought experiment shows a situation that undermines a crucial inference. It does not challenge the premises the way the first type of negative thought experiment does. In terms of logic, this class of thought experiment aims to show invalidity (i.e. the premises may be true but the conclusion does not follow from them). Consider the kind of thought experiment we would present to undermine Lucretius's thought experiment for infinite space. Imagine that we are two-dimensional bugs living on a sphere. Every time we throw a spear at an alleged edge of space it passes through or bounces back because of some barrier. In either case we would agree with Lucretius that this is not the edge of space. However, the inference that space is infinite would be clearly false, since the sphere is finite. Poincaré's disk people example works in a similar way. The measuring rods of the disk people shrink as they move toward the edge of their space, so that they might even come to mistakenly think that they live in an infinite space. If they threw spears, those spears would behave just as Lucretius says. But the disk is finite.

The third type of negative thought experiment—the one I am chiefly concerned with here—is the *Counter Thought Experiment*. The balance of this paper will be devoted to describing them and trying to determine some of their main properties. As examples, I will discuss three:

Galileo against the Aristotelians (principle of relativity)
Mach against Newton (absolute space)
Dennett against Jackson (physicalism)

One of the most interesting features about counter thought experiments is that they are not readily understood in terms of the logic of argument; that is, they are not about validity or soundness. They are directed at a given thought experiment, but they challenge neither the premises nor the concluding inference. Instead, counter thought experiments deny the phenomena of the initial thought experiments.

Boundaries of an experiment

Experimenters do a great many things. They set up their equipment; they let it run and see what happens; they measure;

they calculate; they interpret; and they draw some conclusions which they publish. It is not easy to draw the boundaries of an experiment. The distinction between theory and observation, for instance, is fuzzy at best and the case has been well-made that observations everywhere are theory-laden. I readily grant this, but wish to focus on something a bit more mundane, a distinction between experiment in the broad sense and in the narrow sense.

In the narrow sense, an experiment includes the set up and the observation (which may be highly theory-laden). In the broad sense, the experiment includes background assumptions and initial theorizing, the setup, observation, additional theorizing, calculating, and drawing the final conclusion. It is this final result, with an account of how it was obtained, that we read in a journal. I doubt there could be a sharp distinction drawn between narrow and broad experiment. And what goes for real experiments goes for thought experiments, too. But there is a rough and ready distinction. The narrow part is the phenomenon, it is what we see. We could put this in a simple schematic way:

Theory & Background → Phenomenon → Result

The narrow sense of experiment (whether real experiment or thought experiment) is what we observe, the phenomenon, the middle of the schema. The broad sense includes the whole thing from theory and background assumptions to the final result.

Looking at the simple schema, it is obvious that challenges could come at different points. (NB. The arrows just mean 'is followed by', but for some purposes they might be taken loosely as deductive or inductive implications.) One could challenge the assumptions that played a role in the set up, that is, the theoretical and other background assumptions that went into it. This is what the first class of negative thought experiments do; they attack the premises. One could also challenge the inference to the final result from what went before. This is what the second class of negative thought experiments do; they attack the alleged validity. But obviously, there is also a third way, one could challenge the phenomenon of the thought experiment; that is, one could claim that the phenomenon does not occur, or that what is observed is quite different from what was initially claimed. Let's illustrate this with some examples of each of these types of challenges. In the first set of examples to follow, the phenomena is never at issue; they are *not* examples of Counter Thought Experiments. I include them to provide a useful contrast.

Lucretius, Searle, Thompson

Two thought experiments undermine Lucretius. They both work the same way; they accept the background beliefs of Lucretius, they accept the way the thought experiment is set up, and they accept the observations, as described. They reject the conclusion. Both provide a situation where the background and the phenomenon are as Lucretius wants, but the conclusion of infinite space is false. As I said earlier, it's similar to the way one might show a deductive argument to be invalid: provide an interpretation in which the premisses are true but the conclusion is not.

I mentioned two examples above: bugs on a sphere would not encounter an edge to their space, but their universe is finite, nevertheless. The example illustrates the distinction between infinite and unbounded, a distinction that Lucretius and others would not easily recognize until the rise of non-Euclidean geometry and modern topology.

The second example, only briefly mentioned, is more complex, but also instructive. Poincaré asks us to imagine three-dimensional beings like us inside a finite sphere. It is easier to switch this (as is commonly done) to two-dimensional beings living on a finite, flat disk. The peculiar thing about their world is that there is a force, a bit like heat, that makes all objects expand or contract as they move around the disk. The crucial thing is that *all* objects undergo this contraction as they move toward the edge, so it is utterly unobservable to the inhabitants.

The disk has a radius R and objects contract as they move toward the edge in proportion to $(R^2-r^2)/R^2$ (where r is the distance from the centre). So, if an object has length L at the centre, then its length at a distance r from the centre is $L \times (R^2-r^2)/R^2$. At the edge it shrinks to zero. These distances are as measured in the so-called embedding space, the Euclidean space in which we imagine both the disk and ourselves (with our god's eye view) to be located. If the two-dimensional beings measured their universe, they would find that it took infinitely many lengths of their measuring rods to get to the edge, so they might reasonably conclude that they lived in an infinite universe.

The original point of Poincaré's example had nothing to do with Lucretius. He wanted to show something important about how choices are made when we try to establish the geometry of our universe. Poincaré's disk people would find that the sum of the interior angles does not equal 180 degrees, as in Euclidean geometry, but rather would find that the sum is less than 180 degrees. So they might reasonably conclude that they live in a Lobachevskian (or hyperbolic) universe. The consequence for the status of geometry, according to Poincaré's is this: It is a conventional choice, based on practical considerations, influenced by experience but not determined by it.

I'm only using the first part of Poincaré's thought experiment, the part that involves the experience of the disk people who shrink along with their measuring rods as they move to the edge of their universe. If they threw a spear, it would sail through any alleged edge of space. But, clearly, the inference they might be tempted to make, namely, that space is infinite, is wrong.

These two negative thought experiments both accept the set up and the phenomenon of the initial Lucretius thought experiment (i.e., we never come to an edge). They deny that the conclusion (infinite space), follows from this. They (in effect), attack the validity of Lucretius's thought experiment.

John Searle and Judith Thompson have produced two of the most famous thought experiments of recent times. Searle's Chinese room thought experiment imagines a person in room with an input slot and an output slot through which pass messages in Chinese writing. The person inside has a book that tells him, on a given input, what the output should be. This set up would pass the Turing test; that is, it can think, according to the view of AI (strong artificial intelligence). But, Searle claims, obviously, the person doesn't understand Chinese at all.

There have been numerous challenges to this thought experiment, but none attack the phenomenon. No one denies that there could be such a man in a room receiving and sending messages in Chinese in accord with a book of instructions, yet not understanding the Chinese messages at all. Such an attack, were one to exist, would be what I call a counter thought experiment. Instead, the challenge is usually that Searle has drawn the wrong inference. One claim, for instance, is that it is not the man in the room that passes the Turing test, rather, it is the whole system: room, instruction book, and man. And it is the whole system that understands Chinese, not any part of it, such as the man alone.

Thompson imagined a person hooked involuntarily to a famous violinist (who happened to be unconscious of what happened). The violinist is innocent and has a right to life. The healing process will take nine months, connected all the while; and he would die without being connected for this duration. Though it might be a very generous act to donate one's life for this period, it is surely not morally required. Abortion is analogous to this and so, abortion is morally permissible, even thought (for the sake of the argument), it is granted that the fetus is an innocent person with a right to life. Thompson's thought experiment helps us to make a conceptual distinction: 'right to life' does not equal 'right to what is needed to sustain life.' The violinist/fetus has the former, but not the latter, which is why abortion is morally permissible.

This thought experiment has been repeatedly criticized and rejected, but attacks have not attempted to deny the possibility of actually finding one's self hooked to a violinist who must remain connected for nine months in order to survive. In short, the phenomenon of the thought experiment is not challenged.

In each of these cases, Lucretius, Searle, and Thompson, the challenge has not been directed against the phenomenon, but rather at some other point in the thought experiment. The phenomena in each of them has been undisputed. I turn now to the interesting cases where this is not so, that is, to cases when the phenomenon of a thought experiment has been the focus of attack.

Counter Thought Experiments

As I mentioned above, there are three examples of counter thought experiments that I want to discuss at length. First, Galileo denies Aristotelian thought experiment concerning moving earth; second, Mach denies Newton's thought experiment concerning absolute space; and third, Dennett denies Jackson's thought experiment concerning physicalism.

1. Galileo against Aristotle

From the time of Aristotle through the middle ages, there was a commonly used Aristotelian thought experiment to show the earth could not move. Suppose, on the contrary, that the earth does move. Then a dropped object would fall behind us as we move along; it would not fall straight down to our feet. But, as a matter of fact, it does fall straight down. Thus, the supposition must be false; the earth does not move.

Galileo put forward a counter thought experiment. Not only is it a gem in its own right, but it played a huge role in the development of physics. It established, in effect, the principle of relativity (often now called Galilean relativity).

James Robert Brown

Shut yourself up with some friend in the main cabin below decks on some large ship, and have with you there some flies, butterflies and other small flying animals. Have a large bowl of water with some fish in it; hang up a bottle that empties drop by drop into a wide vessel beneath it. With the ship standing still, observe carefully how the little animals fly with equal speed to all sides of the cabin. The fish swim indifferently in all directions; the drop falls into the vessel beneath; and, in throwing something to your friend, you need throw no more strongly in one direction than another, the distances being equal; jumping with your feet together, you pass equal spaces in every direction. When you have observed all these things carefully (though there is no doubt that when the ship is standing still everything must happen in this way), have the ship proceed with any speed you like, so long as the motion is uniform and not fluctuating this way and that. You will discover not the least change in all the effects named, nor could you tell from any of them whether the ship was moving or standing still. (*Dialogo* 186f)

Galileo's thought experiment denies the phenomenon of the Aristotelian thought experiment. If the earth were moving, a dropped object would land at our feet, not behind us at the initial thought experiment declared. Of course, this does not establish that the earth is indeed moving. The Aristotelian conclusion of a stationary earth might be true. But it does show that the empirical evidence we have is compatible with a moving and with a stationary earth. The Aristotelian thought experiment fails, since things would look the same regardless. This is a counter thought experiment. It denies the phenomenon (that objects would fall behind a moving earth), in the original thought experiment.

2. Mach against Newton

The background to Newton's famous thought experiment concerns rival understandings of the nature of space. Newton's absolutism (often called 'substantivalism'), is the view that space is a substance that exists without depending on anything else. It is the source of inertia. Relationalism is the standard rival view: space is a system of relations. If there were no bodies, there would be no space.

Leibniz, of course, is the prime representative, though he expressed his views most clearly only after Newton's thought experiment.

I hold space to be something merely relative, as time is; that I hold it to be an order of coexistences, as time is an order of successions. For space denotes, in terms of possibility, an order of things which exist at the same time, considered as existing together; without enquiring into their manner of existing. And when many things are seen together, one perceives that order of things among themselves. (Leibniz, *Leibniz-Clarke Correspondence*, 25f)

A 'Leibniz shift' would be moving the whole universe to the right, or mirror reflecting it, etc. But such a thing, Leibniz claimed, is impossible.

I say then, that if space was an absolute being, there would something happen for which it would be impossible there should be a sufficient reason. Which is against my axiom. And I prove it thus. Space is something absolutely uniform; and, without the things placed in it, one point of space does not absolutely differ in any respect whatsoever from another point of space. Now from hence it follows, (supposing space to be something in itself, besides the order of bodies among themselves,) that 'tis impossible there should be a reason, why God, preserving the same situations of bodies among themselves, should have placed them in space after one certain particular manner, and not otherwise; why every thing was not placed the quite contrary way, for instance, by changing East into West. But if space is nothing else, but that order or relation; and is nothing at all without bodies, but the possibility of placing them; then those two states, the one such as it now is, the other supposed to be the quite contrary way, would not at all differ from one another. Their difference therefore is only to be found in our chimerical supposition of the reality of space in itself. But in truth the one would exactly be the same thing as the other, they being absolutely indiscernible; and consequently there is no room to enquire after a reason of the preference of the one to the other. (*ibid*. 26)

Newton expressed his absolutism in the following much quoted passage:

Absolute space, in its own nature, without relation to anything external, remains always similar and immovable. Relative space is some movable dimension or measure of the absolute spaces; which our senses determine by its position to bodies ...(*Principia*, 6)

The bucket thought experiment, surely one of the most famous thought experiments ever, is described as follows:

> ... the surface of the water will at first be flat, as before the bucket began to move; but after that, the bucket by gradually communicating its motion to the water, will make it begin to revolve, and recede little by little from the centre, and ascend up the sides of the bucket, forming itself into a concave figure (as I have experienced), and the swifter the motion becomes, the higher will the water rise, till at last, performing its revolutions in the same time with the vessel, it becomes relatively at rest in it. (*ibid.* 10)

In stage I, the surface of the water is flat and the water is at rest with respect to the bucket. In stage II, the water rotating with respect to the bucket. In stage III, the water at rest with respect to the bucket, but the surface is concave. What's the difference between I and III? Newton offers what seems like the best explanation (and possibly the only one): the water (as well as the bucket), is rotating with respect to space itself.

Of course, the bucket experiment can easily be performed as a real experiment, which presents a problem. The rest of the universe is obviously present around us, something to which a relationalist might appeal. Thus, a second thought experiment is needed and is perhaps even more effective than the bucket. Newton imagines two globes in otherwise empty space.

> It is indeed a matter of great difficulty to discover ... the true motions of particular bodies from the apparent; because the parts of that immovable space ... by no means come under the observation of our senses. Yet the thing is not altogether desperate ... For instance, if two globes, kept at a distance one from the other by means of a cord that connects them, were revolved around their common centre of gravity, we might, from the tension of the cord, discover the endeavour of the globes to recede from the axis of their motion ... And thus we might find both the quantity and the determination of this circular motion, even in an immense vacuum, where there was nothing external or sensible with which the globes could be compared. But now, if in that space some remote bodies were placed that kept always position one to another, as the fixed stars do in our regions, we could not indeed determine from the relative translation of the globes among those bodies, whether the motion did belong to the globes or to the bodies. But if we observed the cord, and found

that its tension was that very tension which the motions of the globes required, we might conclude the motion to be in the globes, and the bodies to be at rest ... (*ibid.*, 12)

Leibniz had no reply to this. Position and velocity are not observable, but acceleration is. The bucket and the rotating globes seemed to establish absolutism. The first serious challenge to Newton on rotation (i.e., accelerating bodies) was from Berkeley and Mach.

Mach begins his challenge to Newton with an assertion of empiricism and a new outlook on inertia. In standard Newtonian mechanics, for instance, we explain the flattening of the earth's poles and bulging of the equator in terms of the earth's rotation. And we presume that if instead of the earth rotating, the stars rotated around the earth, then the bulging of the equator would not happen. Mach takes this to be a serious mistake and that inertial forces ought to arise equally either way. This is an expression of what has come to be known as 'Mach's Principle'. With this empiricist-inspired principle in the background, we come to Mach's counter thought experiment in his *Science of Mechanics*:

Newton's experiment with the rotating water bucket simply informs us that the relative rotation of water with respect to the sides of the vessel produces *no* noticeable centrifugal forces, but that such forces *are* produced by its relative rotation with respect to the mass of the earth and the other celestial bodies. No one is competent to say how the experiment would turn out if the sides of the vessel increased in thickness and mass till they were ultimately several leagues thick. (Mach 1960, 284)

Mach's strategy is rather clear. He proposes a new theory: The source of inertia is not space, but rather is very large amounts of mass. He rejects Newton's bucket and two spheres thought experiments in the narrow sense, that is, he denies the phenomena that Newton claimed would be observed. The water in a rotating bucket with very think walls would not climb the wall of the bucket. And in an empty universe the two balls would not act the way Newton says, but would instead move together because of the tension in the cord connecting them. Mach does not literally assert these things, I am taking a liberty. He merely remarks, somewhat rhetorically, 'who could say what would happen?' But the point is perfectly clear: The scenarios I described are as plausible as Newton's. These are counter thought experiments, they deny the phenomena of the initial thought experiments.

165

3. Dennett against Jackson

Qualia are the subjective aspects of experience, feelings of hunger, pleasure, anger, and sensations of colour, smell, and so on. They are accessible to introspection. (One quale, many qualia.) The status of qualia is central to the mind-body problem. Physicalists claim that there is nothing over and above physical facts. So, qualia present some sort of challenge. Are qualia different? Can they be reduced to the physical, or perhaps eliminated? If not, then physicalism would seem to be wrong.

Frank Jackson is a long-time champion of qualia. He produced a famous thought experiment that has been much discussed for more than two decades.

> Mary is a brilliant scientist who is, for whatever reason, forced to investigate the world from a black and white room via a black and white television monitor. She specialises in the neurophysiology of vision and acquires, let us suppose, all the physical information there is to obtain about what goes on when we see ripe tomatoes, or the sky, and use terms like 'red', 'blue', and so on. She discovers, for example, just which wave-length combinations from the sky stimulate the retina, and exactly how this produces via the central nervous system the contraction of the vocal chords and expulsion of air from the lungs that results in the uttering of the sentence 'The sky is blue.' (It can hardly be denied that it is in principle possible to obtain all this physical information from black and white television, otherwise the Open University would of necessity need to use colour television.)

> What will happen when Mary is released from her black and white room or is given a colour television monitor? Will she learn anything or not? It seems just obvious that she will learn something about the world and our visual experience of it. But then it is inescapable that her previous knowledge was incomplete. But she had all the physical information. Ergo there is more to have than that, and Physicalism is false.

> Clearly the same style of Knowledge argument could be deployed for taste, hearing, the bodily sensations and generally speaking for the various mental states which are said to have (as it is variously put) raw feels, phenomenal features or qualia. The conclusion in each case is that the qualia are left out of the physicalist story. And the polemical strength of the Knowledge argument is that it is so hard to deny the central claim that one

can have all the physical information without having all the information there is to have. (Jackson 1982, 130)

Here is the all important knowledge argument that comes from the thought experiment.

1. Mary knows all the physical facts about colour perception.
2. She has learned these facts having only black and while experiences.
3. When she experiences colour for the first time, she learns something new.

Therefore, some facts about colour are not physical facts.

Before getting to Dennett's thought experiment, let me first take note of his outright rejection of this or indeed of any thought experiment.

Like a good thought experiment, its point is immediately evident even to the uninitiated. In fact it is a bad thought experiment, an intuition pump that actually encourages us to misunderstand its premises. (Dennett 1991, 398).

Thought experiments depend on folk concepts; they are inherently conservative. We should expect very counter intuitive results in real science, so violating our intuitions is to be expected (Dennett 2005, 128f)

Dennett raises an important point about intuitive concepts. But his dismissal of thought experiments because they make use of them is quite unjustified. So called folk concepts—whether they are used in thought experiments or not—can and often do lead to revolutionary results. I'll take a moment to bludgeon readers with examples.

- Galileo's thought experiment on free fall led to a new mechanics.
- Poincaré's disk thought experiment lead to very rich model of non-Euclidean geometry
- Einstein's elevator thought experiment lead to the principle of equivalence which is central to General Relativity.
- Thompson's violinist thought experiment leads to new view of the morality of abortion.

It's not just thought experiment where this happens; let me mention a few other examples.

- Arithmetic deals with very simple (folk) concepts of addition, multiplication, and division. With these we can define 'prime number' and easily prove a very profound theorem that there are infinitely many primes.
- From the very simple concepts of arithmetic we can (step by common sense step) go on to establish the remarkable result by Gödel of the incompleteness of any set of axioms for arithmetic.
- Turing computability can readily be seen as nothing more than the elaboration of common sense concepts of rule-governed calculation, but it leads to the unexpected result that there are uncomputable functions.

I certainly don't want to say that everything is at bottom based on folk concepts. 'Isospin', 'superego', 'magnetic field', and many other important notions are certainly not commonsense ideas at all, but must be introduced in some conjectural fashion. But it's a mistake to think that starting with common sense must end in common sense. Dennett's dismissal of "intuition pumps" is quite misguided. Fortunately, Dennett condescends to play the thought experiment game, anyway, and he does so with considerable success.

Dennett's attack on Jackson is in the form of the following counter thought experiment. The setup is the same as Jackson's, but the scenario is quite different.

And so, one day, Mary's captors decided it was time for her to see colours. As a trick, they prepared a bright blue banana to present as her first colour experience ever. Mary took one look and said 'Hey! You tried to trick me! Bananas are yellow, but this one is blue!' Her captors were dumbfounded. How did she do it? 'Simple,' she replied, 'you have to remember that I know everything—absolutely everything—that could ever be known about the physical causes and effects of color vision. So of course before you brought the banana in, I had already written down, in exquisite detail, exactly what physical impressions a yellow object or a blue object (or a green object, etc.) would make on my nervous system. So I already knew exactly what thoughts I would have (because, after all, the mere disposition to think about this or that is not one of your famous qualia, is it?). I was not in the slightest surprised by my experience of blue (what surprised me was that you would try such as second-rate trick on me). I realise that it is hard for you to imagine that I could know so much about my reactive dispositions that the way blue affected me came as no surprise. Of course it's hard for you to imagine. It's hard for anyone to imagine the consequences of someone knowing absolutely everything physical about anything! (Dennett 1991, 399f)

It should be quite clear at this point what is happening. Jackson's thought experiment has the following structure: There is a set up: Mary is in black and white room, where she learns all physical facts about perception. Next comes the phenomenon: When Mary first encounters colours, she learns something new. Finally, the result of the thought experiment, i.e., the conclusion drawn: Some facts about perception are not physical, and so, physicalism is wrong. A counter thought experiment would accept the set up, but challenge the phenomenon, which is exactly what Dennett does.

I hope the general conclusion I wish to draw from these three examples is evident: Dennett = Mach = Galileo. That is, the structure of Dennett's thought experiment is the same as Mach's and Galileo's. They are all counter thought experiments. The challenge for Dennett was not: given the thought experiment we should resist the anti-physicalist conclusion (i.e., he is not against the broad thought experiment). Rather, Dennett's challenge is that the narrow thought experiment is faulty; the phenomenon is not as Jackson claims it would be. Mary would not learn anything new. Dennett, Mach, and Galileo each deny the phenomena of the initial thought experiments.

Alternative Challenges

The Newton and Jackson thought experiments might also be challenged in the broad sense (i.e., by accepting the phenomena of the thought experiment but offering a different explanation). The challenges would not be in the form of a counter thought experiment, possibly not in the form of a thought experiment at all. I'll briefly mention two examples, just for the sake of comparison.

Contra Newton, Larry Sklar introduced his notion of *absolute acceleration* (Sklar, 1976). An object or system of objects, such as the two spheres connected by a cord, might have this property. When it does, there will, for instance, be a tension in the cord joining the two spheres. They are not rotating with respect to anything, they simply have this property of absolute acceleration. It's quite bizarre, but if one thinks of quantum mechanical spin, then one gets the idea. The spin of an electron is 'intrinsic,' it cannot be transformed away in any coordinate frame. Sklar's account, unlike Mach's, does not challenge the phenomenon of Newton's thought experiment; it offers a different explanation.

Contra Jackson, David Lewis proposed the 'ability hypothesis'. (Lewis 1983, 1988) It is related to the distinction between knowing how, not knowing that. One might know absolutely every fact about a bicycle, yet not know how to ride. If one learns how to ride, one is not learning a new fact or acquiring new propositional knowledge, but rather one is acquiring a skill, a new ability. When Mary leaves the laboratory and experiences red things for the first time, she is similarly learning a skill, not learning a new fact. Lewis's account, unlike Dennett's, does not deny the phenomenon of Jackson's thought experiment, but rather undermines Jackson's knowledge argument by interpreting the phenomenon of the thought experiment in a different way.

Evaluation in these cases takes the form: Who offers the best explanation or interpretation of the phenomena in the thought experiment? Sklar and Lewis do not deny the phenomena of Newton's and Jackson's thought experiments. Rather they challenge the inference drawn after accepting the phenomena.

When Does a Counter Thought Experiment Work?

The main aim of this paper is the modest one of pointing out the existence of a distinct class of counter thought experiments. But once we accept the existence of counter thought experiments and get some idea of how they work, the inevitable questions to ask are: when do they work well?, and when do they fail? What follows is but a superficial start at addressing these questions.

Clearly, a counter thought experiment will work only when it can plausibly deny the phenomenon of the original thought experiment. I don't think anyone could reasonably hope to deny the phenomena in, say, Searle's Chinese room thought experiment. Everyone is ready to allow that a person could be in a room with Chinese characters taken in that are compared by a person inside with those in a book that tell her which Chinese characters to put out. Challenges to Searle's thought experiment have all been aimed at the inference that he drew from the phenomenon. Similarly, there is no point in rejecting the phenomenon in Lucretius infinite space thought experiment, since it merely involves throwing a spear. To do so would involve a degree of scepticism that goes well beyond the case at hand. The same could be said of Thompson's violinist. Of course, we could wake up with an unconscious and very ill violinist hooked up to ourselves such that he will be cured if and only if he remains connected for nine months. There could be no plausibility to denying that such things could happen. Challenges to these thought experiments must be aimed at the various conclusions the thought experimenter draws from the directly observed part of the thought experiments.

This should be uncontentious, but there are those who would disagree. Peter Geach, for instance, takes moral rules to be devine commands and he holds that God would not allow genuine moral dilemmas to exist, since we would then have to choose between different devine commands, which he takes to be absurd. Geach imagines someone saying: ' "But suppose circumstances are such that observance of one Divine law, say the law against lying, involves breach of some other absolute Divine prohibition" ' Geach then replies:

—If God is rational, he does not command the impossible; if God governs all events by his providence, he can see to it that circumstances in which a man is inculpably faced by a choice between forbidden acts do not occur. Of course such circumstances (with the clause 'and there is no way out' written into their description) are consistently describable; but God's providence could ensure that they do not in fact arise. Contrary to what unbelievers often say, belief in the existence of God does make a difference to what one expects to happen. (Geach 1969, 128)

The upshot, according to Geach, is that perfectly consistent thought experiments might still be illegitimate and hence the phenomenon not exist, because God would not allow it to happen. I mention this outlook in passing to further illustrate the range of possible opinion on thought experiments. It is not one I think we should seriously consider.

Here are some things that seem to matter when evaluating a counter thought experiment. They are all rather obvious.

- How reliable is the initial thought experiment in the narrow sense (i.e., would the phenomenon occur)?
- How strong is the assumed background to the thought experiment?
- How similar is the phenomenon of the thought experiment to things we know and trust?

James Robert Brown

- How plausible is the phenomenon of the counter thought experiment?

  - How absolutely plausible?
  - How relatively plausible?

The last of these is probably the key question, but let's flesh them all out a bit by considering our three examples.

In making the case for the Aristoteleans, we might note what happens when we throw some litter out the car window—it falls far behind as we move along. (I hope introducing a car is a harmless anachronism, and I can assure readers I am not a litterbug.) The case for Galileo could be based on our experience of tossing things around in a moving car, aeroplane, etc. Motion seems to have no effect. By analogy, if the whole earth were moving, our experience should be as in a car, plane, etc.

In making the case for Newton, we might note that we have often seen water climb rotating buckets and we have felt the tension in a string holding a rock that is spinning around us. The two globes thought experiment assumes they would act the same in empty space, which seems very plausible. Mach, on the other hand, proposes a new theory: mass is the cause of inertial motion. There is no empirical evidence for this; it's motivated by his rather strict empiricism. (Remember the fate of Mach's empiricist-inspired anti-atomism.) However, it would seem that acceleration should be on par with position and velocity—*relative*, otherwise not detectable.

Given that Mach's account is possible, he undermines to some extent the degree of belief we had in the phenomenon of Newton's thought experiment, i.e., that the cord's tension would be maintained. But Mach's counter thought experiment is certainly not as plausible as Newton's. Consequently, it is a weak attack on Newton's thought experiment, and hence, a weak attack on absolute space.

The Case for Jackson might begin by noting that in general, mental things don't seem like physical things, and more specifically, when people, for instance, acquire eye-sight late in life, they appear to learn something new. Mary the colour scientist seems like an extreme case of this; hence, Jackson's narrative appears initially plausible. But we don't know anyone who knows everything about anything, much less all the physical facts. So, the analogy with things we already know is very weak. There is a superficial similarity to Plato's cave, but with vastly greater assumptions.

172

The case for Dennett might begin with noting that for various philosophical reasons, physicalism seems right (i.e., problems with dualism, etc.). And, to repeat, we have no idea what it would be like for someone to know all the physical facts. As a story, Dennett's narrative about Mary seems coherent and intelligible. It would appear then that Dennett's counter thought experiment is just as plausible as Jackson's, even though neither is very plausible in its own right. Dennett is quite aware of this:

> My variant was intended to bring out the fact that, absent any persuasive argument that this could not be the way Mary would respond, my telling of the tale had the same status as Jackson's: two little fantasies pulling in opposite directions, neither with any demonstrated authority. (Dennett 2005, 105)

Thus, Jackson is neutralized, if not refuted, and the Mary thought experiment is a failure.

Comparatively, I would say that Galileo is completely triumphant; the Aristotelian thought experiment is destroyed. Newton is slightly weakened but not seriously damaged; Mach is an alternative, but it is nowhere near as plausible. Jackson is nullified, since Dennett's alternative story is equally plausible. It's a tie, as far as the thought experiments go, which probably leaves Dennett the winner in this particular battle.

These evaluations, of course, are very rough and open to objection. They are only preliminary and should not be taken too seriously. They merely illustrate the kinds of consideration involved. My main aim is to determine how counter thought experiments work in general, not to evaluate particular instances.

I do, however, wish to explore the comparative nature of the thought experiment-counter thought experiment pair by briefly examining a simple proposal. Seeing its shortcomings will, I hope, stimulate some interest in others in the further investigation of counter thought experiments.

A Ratio Test

For quite some time, it has been common to think of the evaluation of scientific theories as taking place comparatively. Kuhn's paradigms and Lakatos's research programmes are evaluated (at least in part), by comparing them with rivals. Much the same can be said of counter thought experiments. There are, however, important differences. Comparative theory evaluation is usually

James Robert Brown

over the long haul and it is the whole theory/paradigm/ programme that is being compared. By contrast, thought experiments and counter thought experiments go head to head and the evaluation is direct and immediate.

In trying to capture this comparative aspect, we might try the following *Ratio Test*, as I shall call it. Assign a probability to the phenomenon of a thought experiment, given the thought experiment set up. I should readily admit and even stress that this not intended to be realistic; I doubt these things can be quantified. But it might shed a little light on the structure of counter thought experiments.

Let *initial phen* = the phenomenon in the original thought experiment (e.g., action of Newton's two spheres in empty space, or Mary learning something new when she leaves the laboratory), and let *counter phen* = the phenomenon in the counter thought experiment (e.g., action of the two spheres according to Mach, or the actions of Mary in Dennett's thought experiment). Assign probabilities to these, e.g., Prob(Mary learns something new) = r. Probability here is meant to be something like degree of belief.

It would seem that a counter thought experiment is successful, if: Prob(counter phen)/Prob(initial phen) >1, and is not successful, if: Prob(counter phen)/Prob(initial phen) << 1.

Why does the second claim not use ≤, which would be a simple denial of the first? The reason has to do with a complication I have not mentioned, but will soon be obvious.

Presenting a counter thought experiment is perhaps like the defence presenting an alternative account of the facts of a legal case. The prosecution must make its case 'beyond a reasonable doubt'. The defence need not match that high standard, but need only make a case for a slightly plausible alternative. This asymmetry in standards will upset the ratio test, or at least would greatly complicate it. Even if we think the prosecution's story is more likely, the possibility presented by the defence is enough to undermine our initial confidence. In probablilistic terms, the defence can do its job successfully even when its case has a probability well below ½, just as long as the probability isn't too low. Mach's counter thought experiment might plausibly fall in this range. It may not be plausible in its own right, but it could be plausible enough to undermine our initial assessment of Newton's thought experiment.

In general, the range of plausibility of counter thought experiments is great. Some counter thought experiments might be highly compelling in their own right, as was Galileo's. Others might be weak in their own right, but still strong enough to cast doubt on the main account, as Mach's perhaps did to Newton's.

There are also cases where the ratio test might break down badly. This will happen in 'how possible' thought experiments. These are thought experiments that don't try to establish a result concerning how things are, but only try to show how something is possible. Darwin provided examples in discussion of the evolution of particular characteristics that seemed problematic. How could the eye evolve or the giraffe acquire a long neck? Darwin's thought experiment would show a possible evolutionary route. It was not intended to be true, only to show that the particular characteristic is not a counter example to the theory of evolution. As long as its probability is not equal to zero, the thought experiment is a success. One of Darwin's foremost early critics, Fleming Jenkins, produced counter thought experiments (involving 'blended inheritance'), that aimed to show the evolutionary account Darwin provided is not possible. In other words, he constructed counter thought experiments with probability virtually equal to one. (See Lennox 1991 for an account of Darwin and Jenkins.)

The ratio test is quite inappropriate in cases such as these, since almost inevitably Prob(counter phen)/Prob(initial phen) >> 1. This will happen even when the counter thought experiment is only moderately plausible, since the initial thought experiment is only meant to show possibility, not likelihood. The ratio test is at best a first stab; it is certainly not adequate as it stands. Counter thought experiments that aim to undermine 'how possible' thought experiments will have to be evaluated some other way.

Concluding Remarks

There is an interesting class of negative thought experiments, which I have called *Counter Thought Experiments*. Galileo against the Aristotelians on the motion of the earth, Mach against Newton on absolute space, and Dennett against Jackson on physicalism are instances. Evaluation of these counter thought experiments seems to be essentially comparative. A simple proposal, a ratio test, works reasonably well in some cases, but it will certainly need supplementing, since it flounders on 'how possible' cases.

Is it possible to give a general account—perhaps quite different from the one I have sketched—of what makes a counter thought experiment effective? This is wholly unexplored territory, but

175

James Robert Brown

definitely worth further investigation, as are all areas of the remarkable topic of thought experiments.

Bibliography

Brown, J.R. (1986) 'Thought Experiments Since the Scientific Revolution', *International Studies in the Philosophy of Science*, Vol I, no 1. 1986.

Brown, J.R. (1991) *Laboratory of the Mind: Thought Experiments in the Natural Sciences*, London: Routledge.

Brown, J.R. (1999) *Philosophy of Mathematics: An Introduction to the World of Proofs and Pictures*, London: Routledge.

Brown, J.R. (2003a) 'Peeking Into Plato's Heaven', *Philosophy of Science*, vol. 71, 1126–1138.

Dennett, D. (1991) *Consciousness Explained*, New York: Little Brown.

Dennett, D. (2005) *Sweet Dreams*, Cambridge, MA: MIT Press.

Galileo (*Dialogo*), *Dialogue Concerning the Two Chief World Systems* (Trans from the *Dialogo* by S. Drake), second revised edition, Berkeley: University of California Press, 1967.

Galileo (*Discoursi*), *Two New Sciences*, (Trans from the *Discoursi* by S. Drake) Madison: University of Wisconsin Press, 1974.

Geach, P. (1969) *God and the Soul*, London: Routledge and Kegan Paul.

Horowitz, T. and G. Massey (eds.) (1991) *Thought Experiments in Science and Philosophy*, Savage MD: Rowman and Littlefield.

Jackson, F. (1982) 'Epiphenomenal Qualia', *Philosophical Quarterly*, Vol. 32, No. 127, 127–136.

Kuhn, T. (1964) 'A Function for Thought Experiments', reprinted in Kuhn, *The Essential Tension*, Chicago: University of Chicago Press, 1977.

Leibniz, G. (1956) *The Leibniz-Clarke Correspondence*, H. G. Alexander (ed.) Manchester: Manchester University Press.

Lennox, James G. (1991). 'Darwinian Thought Experiments: A Function for Just-So Stories', in Horowitz and Massey (1991), 223–245.

Lewis, D. 1983. Postscript to 'Mad Pain and Martian Pain'. In his *Philosophical Papers*, Vol. 1. New York: Oxford University Press, 1983, 130–32.

Lewis, D. 1988. 'What Experience Teaches'. In *Proceedings of the Russellian Society*. Sydney: University of Sydney, 1988. Reprinted in *Mind and Cognition*, W. Lycan (ed.), Oxford: Blackwell, 1990, 499–518.

Lucretius, *De Rerum Natura*, Cambridge,Ma, Loeb Library.

Mach, E. (1960) *The Science of Mechanics*, (Trans by J. McCormack), sixth edition, LaSalle Illinois: Open Court.

Mach, E. (1976) 'On Thought Experiments', in *Knowledge and Error*, Dordrecht: Reidel.

Newton, I. (Principia) *Mathematical Principles of Natural Philosophy* F. Cajori (trans.), Berkeley: University of California Press

Norton, J. (1991) 'Thought Experiments in Einstein's Work', in Horowitz and Massey (1991).

Norton, J. (1996) 'Are Thought Experiments Just What You Always Thought?' *Canadian Journal of Philosophy*.

Poincaré, H. (1969) *Science and Hypothesis*, New York: Dover.

Searle, J. (1980) 'Minds, Brains, and Programs', *Behavioral and Brain Sciences* 3, 417–424.

Sklar, L. (1976) *Space, Time, and Spacetime*, Berkeley, University of California Press.

Thompson, J.J. (1971) 'A Defense of Abortion', *Philosophy and Public Affairs*, 1/1 (Fall): 47–66.

Does Physics Answer Metaphysical Questions?[1]

JAMES LADYMAN

1. Introduction

According to logical positivism, so the story goes, metaphysical questions are meaningless, since they do not admit of empirical confirmation or refutation. However, the logical positivists did not in fact reject as meaningless all questions about for example, the structure of space and time. Rather, key figures such as Reichenbach and Schlick believed that scientific theories often presupposed a conceptual framework that was not itself empirically testable, but which was required for the theory as a whole to be empirically testable. For example, the theory of Special Relativity relies upon the simultaneity convention introduced by Einstein that assumes that the one-way speed of light is the same in all directions of space. Hence, the logical positivists accepted an a priori component to physical theories. However, they denied that this a priori component is necessarily true. Whereas for Kant, metaphysics is the a priori science of the necessary structure of rational thought about reality (rather than about things in themselves), the logical positivists were forced by the history of science to accept that the a priori structure of theories could change. Hence, they defended a notion of what Michael Friedman (1999) calls the 'relativised' or the 'constitutive' a priori. Carnap and Reichenbach held that such an a priori framework was conventional, whereas Schlick seems to have been more of a realist and held that the overall relative simplicity of different theories could count as evidence for their truth, notwithstanding the fact that some parts of them are not directly testable. All this is part of the story of how the verification principle came to be abandoned, and how logical positivism transmuted into logical empiricism.

[1] This text is more or less that of a talk of the same title given to the *Royal Institute of Philosophy* on 10th February 2006. The relationship between science and metaphysics is discussed in more detail in Ladyman and Ross (2007).

Yet there is no doubt that the crude view that the positivists banished metaphysics altogether was considerably influential despite the more nuanced history that is now being told. For example, here is John Watkins in 1975:

> The counter-revolution against the logical empiricist philosophy of science seems to have triumphed: I have the impression that it is now almost as widely agreed that metaphysical ideas are important in science as it is that mathematics is. (91)

Watkins celebrates the work of historians of science who showed 'the decisive role that certain metaphysical speculations have played in the advancement of science' (91). Since he wrote it has become ever more widely acknowledged that metaphysical questions are meaningful and important, and metaphysics is now enjoying something of a renaissance. In philosophy of science, the acceptance of metaphysics is due in part to the emergence and dominance of scientific realism. For one definition of metaphysics is that it is the study of the nature of reality beyond the appearances of things, and scientific realism is the position that we ought to believe in the unobservable entities, such as atomic and subatomic particles, fields, and black holes, postulated by our best scientific theories. Unobservable entities are now fundamental to a great deal of science and especially physics, and the theories that describe these unobservable entities clearly purport to describe reality beyond the appearances. Therefore, that metaphysical questions can be answered by physics seems to follow from scientific realism.[2] This is surely a significant general consideration (of which more below), but perhaps a more important part of the explanation of the seriousness with which philosophers of science now take metaphysics is the fact that a number of fundamental metaphysical doctrines are widely perceived to be important parts of the content of physical theories. For example, Special Relativity is thought to imply that all times are real, and quantum mechanics is thought to imply indeterminism. (More of these specific cases below.)

However, other factors have led to metaphysics being regarded as not merely a meaningful and legitimate part of philosophy, but also as an autonomous and non-empirical science. One of these is that philosophers of science began to turn to metaphysics in an attempt

---

[2] Indeed a recent textbook (Couvalis, 1997) on philosophy of science defines scientific realism as the view that metaphysical questions can be answered by science.

to explain the continuity of reference between successive scientific theories that gave competing descriptions of unobservable entities. The causal theory of reference, and ideas such as rigid designation, were in part a response to the challenge to scientific realism raised by the historicism of Kuhn and others. Hence, it began to look as if scientific realism needed to be grounded in metaphysics. Furthermore, there was a general acceptance in analytic philosophy after the heyday of Quineanism that modality could not be eliminated either from philosophy or from the rational reconstruction of science, and so the metaphysics of possible worlds, causation, and laws became central areas of philosophical concern. The philosophers pursuing these issues did not detect any prospect of science being able to settle them, and so nowadays metaphysics is pursued by many who regard it as sovereign.

2. Autonomous versus Naturalised Metaphysics

It is a commonplace among contemporary philosophers that the systematic divide among the great early modern philosophers into rationalists and empiricists is an artefact of the histories of philosophy from the late nineteenth century that whiggishly construed rationalism and empiricism as thesis and antithesis prior to celebrating Kant's cunning synthesis. That said, there is no doubt that we can detect the binary opposition within contemporary philosophy, and as ever no where is it more evident than within metaphysics. Representing the resurgent voice of the analytic metaphysicians here is Jonathan Lowe:

> ... metaphysics goes deeper than any merely empirical science, even physics, because it provides the very framework within which such sciences are conceived and related to one another (2002, vi)

According to him the universally applicable concepts that metaphysics studies include those of identity, necessity, causation, space and time. Metaphysics must say what these concepts are and then address fundamental questions involving them, such as whether causes can have earlier effects. Metaphysics other main job according to Lowe, is to systematise the relations among fundamental metaphysical categories such as things, events, properties, and so on. Leaving aside the latter, we might reasonably

181

ask how metaphysical enquiry ought to proceed and Lowe adopts the familiar methodology of reflecting on our concepts (conceptual analysis).

Frank Jackson (1998) has also recently defended the idea of metaphysics as conceptual analysis. The problem these philosophers have is with explaining why we should think that the products of their activity reveal anything about the deep structure of reality rather than merely telling us about how we think about and categorise reality. Even those fully committed to a conception of metaphysics as the discovery of synthetic a priori truths shy away from invoking a special faculty of rational intuition that delivers such knowledge; rather they usually just get on with their metaphysical projects and leave the matter of explaining the epistemology of metaphysics for another occasion. Ted Sider defends this strategy by pointing out that lack of an epistemological foundation for science and mathematics does not prevent practitioners from getting on with the business of advancing the state of knowledge in these domains (2001, xv). The obvious rejoinder to this, which he does not consider, is that we know a tree by its fruits and mathematics and science have undoubtedly borne fruits of great value; pure metaphysics has not achieved anything comparable, if it has achieved anything at all.

Furthermore, naturalists have good positive reasons to be sceptical about a priori metaphysics. There is no reason to believe that human beings naturally endowed cognitive capacities extend to the provision of reliable information about the fundamental structure of reality.[3] So the only kind of metaphysical enquiry that ought to be taken seriously, if any, is naturalised metaphysics. (Of course the naturalist concedes that if a metaphysical system is internally consistent then it can be rejected a priori.)

Lowe has two arguments against this:

(i) '... to the extent that a wholly naturalistic and evolutionary conception of human beings seems to threaten the very possibility of metaphysical knowledge, it equally threatens the very possibility of *scientific* knowledge ...' (6) Since natural selection cannot explain how natural scientific knowledge is possible so the fact that it cannot explain how metaphysical knowledge is possible gives us no reason to suppose that such knowledge is not possible.

(ii) Naturalism depends upon metaphysical assumptions.

---

[3] For more on this see Ladyman and Ross (2007), 1.1 and 1.2.

Does Physics Answer Metaphysical Questions?

In response though note that even if it is granted that natural selection cannot explain how natural scientific knowledge is possible, we have plenty of good reasons for thinking that we do have such knowledge. On the other hand, we have no good reasons for thinking that metaphysical knowledge is possible. (Recall the point against Sider above.) It seems that their predilection for a priori argument leads metaphysicians to discuss the status of metaphysics compared to science a priori as if we didn't know a posteriori that the track record of science gives us reason to take it epistemically seriously that we lack in the case of metaphysics. With respect to Lowe's second argument here, it is enough to point out that even if naturalism depends on metaphysical assumptions, the naturalist will argue that the metaphysical assumptions in question are vindicated by the success of science, by contrast with the metaphysical assumptions on which autonomous metaphysics is based which are not vindicated by the success of metaphysics since it can claim no such success.

According to Lowe, it is the job of metaphysics to tell us what is possible, but it may be conceded that which of the possible fundamental structures of reality exists can be answered only with empirical evidence. The problem with this is that philosophers have often regarded as impossible states of affairs that science has come to entertain. For example, metaphysicians confidently pronounced that non-Euclidean geometry is impossible, that it is impossible that there not be deterministic causation, that non-absolute time is impossible, and so on. Furthermore, there is no agreement now among metaphysicians about what is metaphysically possible. For example, some metaphysicians believe that infinitely divisible matter is possible and others that it is not.

In any case, suppose that naturalised metaphysics is the only option. It remains an open question whether it is possible to learn metaphysical lessons from science. Since metaphysics does not give rise to empirical predictions it cannot be directly confirmed by a physical theory. However, metaphysical theories can be incompatible with physical theories, and hence the former can be ruled out or at least undermined by the confirmation and adoption of the latter. For example, Cartesian metaphysics is incompatible with Newtonian physics because the former has it that space is filled with a plenum and that all action is action by contact, whereas the latter involves particles moving in a void subject to the force of gravity that seems to act at a distance. For Watkins, metaphysics is part of the hard core of a Lakatosian research programme. It cannot be directly refuted but if the research programme persistently degenerates then the hard core must eventually be abandoned. Arguably this happened to Aristotelian and Cartesian metaphysics, and then later it happened to the metaphysical framework of classical physics in the face of relativity and quantum physics. As Sider says in a more naturalistic mood: '... in cases of science versus metaphysics, historically the smart money has been on science ...' (2002, 42). (Sider seems to think that metaphysical knowledge can be a priori but that metaphysics is also required to be compatible with current science.)

Let us grant then that the failure of a metaphysical framework to help in the production of good science can count as a reason against it. Is it also plausible that the success of science based on a particular metaphysical framework provides some evidence for the latter? For example, the success of materialism as a research programme in science in the seventeenth and eighteenth centuries was arguably evidence for it. (Of course, the conception of matter integral to materialism, namely that of extended substance, became untenable in the nineteenth and twentieth centuries.)

Katherine Hawley (2006) helpful distinguishes a number of possible positions one might take on the relationship between metaphysics and science. Hawley assumes that it is sometimes at least possible to provide reasons for and against metaphysical claims and that science is not the only source of such reasons; she therefore thinks that 'metaphysical enquiry can be reasonable even where it outruns the scope of science' (454). Purely a priori reasons for metaphysical claims may involve appeals to intuition, explanatory power, simplicity, or perhaps the results of conceptual analysis. As was mentioned above, naturalists will be suspicious of appeals to intuitions not least because intuitions vary so much, and also because even were we agreed in our intuitions, they doubt that there is any reason to believe that our intuitions track the truth about metaphysical matters any more than they do about motion, gravity or the constitution of material objects. When it comes to explanatory power, naturalists must agree that inference to the best explanation is indispensable in science and so the question can only be whether there is any difference between metaphysics and science in respect of explanation. There is an important such difference, namely that what metaphysical hypotheses are supposed to explain are often metaphysician's intuitions or 'common-sense beliefs'. In the light of this, naturalists ought to be very sceptical about the extension of the methodology of inference to the best explanation into metaphysics, and about the worth of pursuing metaphysical enquiry where it goes beyond the scope of scientific enquiry.[4]

However, we can at least agree with Hawley that an outright sceptic about inference to the best explanation like Bas van Frasssen must eschew metaphysics, just as he does scientific realism. Hence, some form of realism about science is a necessary condition for believing that in principle science can answer metaphysical questions. According to Hawley, 'optimism' is the view that 'the involvement of a metaphysical claim in an empirically successful scientific theory provides some reason to think that the claim is true' (456). This idea that in principle science can answer metaphysical questions is the negation of Hawley's 'Radical Pessimism'. She thinks that scientific realism is sufficient for optimism because there is no 'in-principle difference between claims about unobservable 'scientific' entities and 'metaphysical' claims' (459). The idea is that since the theoretical claims of science concerning unobservable entities go beyond the empirical content of the relevant theories, the scientific realist is committed to the idea that the empirical success of theories gives us reason to believe their theoretical content. The onus is on the sceptic about metaphysics to point out some relevant semantic or epistemic difference between the theoretical content and the metaphysical content of theories and there is good reason to think this cannot be done. One of way of thinking about this is in terms of the Quinean notion of the web of belief. Metaphysical hypotheses are as much part of the web as theoretical scientific hypotheses, and since neither can be confirmed or refuted in isolation, in principle they have the same epistemic status.

Let us grant all this for the moment. It follows that the scientific realist must deny radical pessimism and so they must chose between the following two positions:

(1) Metaphysical claims have in fact been established by science.
(2) Although in principle metaphysical claims might be established by science, in fact they have not been.

(Hawley talks not of science establishing metaphysical claims but of it providing some reason to think they are true. However, this is too weak: The sceptic about metaphysics may concede that science gives us some reason to believe a particular metaphysical claim, but deny that the reasons in question are anywhere near compelling.)

---

[4]   Again see Ladyman and Ross (2007), 1.1 and 1.2 and also van Fraassen (2002), chapter 1.

James Ladyman

If it is supposed that scientific realism implies that metaphysical hypotheses have been established by science, and it turns out that metaphysical hypotheses have not in fact been established by science, then by modus tollens, scientific realism must be false. Hence, close attention to the role of metaphysics in physics and to the fate of metaphysical hypotheses in the history of physics may bolster scepticism, not just about metaphysical knowledge, but also about scientific realism. On the other hand, if it is supposed that scientific realism only requires that metaphysical hypotheses can in principle be established by science, but they have not in fact been, then it may be objected that the form of scientific realism that results is too attenuated to really be worth the name. After all, as Hawley says, there is no clear demarcation between a theoretical claim and a metaphysical one, and indeed claims about unobservable entities become empty if nothing is said about their metaphysical status. For example, the claim that light is transmitted as waves in the ether is empty without some account of what the ether is supposed to be.[5]

3. Metaphysical questions that might be thought to be answerable by physics:

Here are some examples:

(i) Are all times real?
(ii) Does time flow?
(iii) Is there a global asymmetry in time in the universe?
(iv) Is time travel possible?
(v) Is the world deterministic[6]?
(vi) Can a cause succeed its effect in time?
(vii) Is there action at a distance or does all causation happen locally in space and time?
(viii) Is the identity of indiscernibles true?
(ix) Do space (and/or time) exist independently of their material contents?
(x) Are space and time discrete or continuous?
(xi) What is the geometry of space(time)?

[5] For more on the status of the metaphysical content of theoretical claims see Ladyman and Ross (2007), 2.2.
[6] Determinism is here understood as a modal thesis: given the state of the world at some time, there is only one way it *could* be at any later (or earlier) time.

(xii) How many dimensions are there to spacetime
(xiii) What is the topology of space(time)?
(xiv) Is the universe finite or infinite?
(xv) Are there parallel universes?

4. The Philosophy of Time

Questions (i) to (iv) all concern time. They are related but nonetheless distinct although often conflated. In relation to (i), call the view that all events are real 'eternalism', the view that only the present is real 'presentism', and the view that all past and present events are real 'cumulative presentism.' (The latter is defended by Michael Tooley (2000) although not under this name). (ii) is often expressed in terms of whether there is temporal passage or objective becoming. Those who believe in the passage of time or objective becoming often also believe that the process of becoming is that of events coming into existence and going out of existence, but this need not be so; to suppose there is becoming, one need only believe that there is some objective feature of the universe associated with the passage of time. Objective becoming could be like a light shining on events as they are briefly 'present', and is therefore compatible with eternalism.[7] On the other hand, both presentism and cumulative presentism entail a positive answer to question (ii), since if events do come into existence, whether or not they then stay existent or pass out of existence, this is enough to constitute objective becoming. Since McTaggart's famous argument for the unreality of time, metaphysicians have often discussed (i) and (ii) in the context of a question in the philosophy of language namely: Is tensed language reducible to tenseless language, or does tensed language have tenseless truth conditions? (*) However, the philosophy of language cannot settle the metaphysical issues as Michael Tooley (2000) argues. So even though the standard opposition is between those who answer 'no' to (i), 'yes' to (ii), and 'no' to (*) on the one hand (the defenders of McTaggart's 'A-series'), and those who answer 'yes' to (i), 'no' to (ii), and 'yes' to (*) (the defenders of McTaggart's 'B-series'), a variety of more nuanced positions are possible. (Note that it is not clear on reflection that the negation of presentism is a necessary condition for the possibility of time travel.)

---

[7] Tim Maudlin (2002) argues that the passage of time may be an objective feature of the universe even if eternalism is true.

James Ladyman

In relation to (iv), clearly if (i) or (ii) are answered positively then that is enough to privilege a particular direction in time. However, eternalism and the denial of objective becoming are compatible with time having a privileged direction, since there could be some feature of the block universe that has a gradient that always points in some particular temporal direction. For example, the entropy of isolated subsystems of the universe, or the universe itself, might always increase in some direction of time. Another well known possible source of temporal direction was proposed by Reichenbach (1956) who argued that temporal asymmetry is grounded in causal asymmetry: in general, correlations between the joint effects of a common cause are screened off by the latter but the joint causes of a common effect are uncorrelated. Although some have claimed that Reichenbach's Principle of the Common Cause is violated, not least by the behaviour of entangled states in quantum mechanics (see for example van Fraassen 1991), such considerations are sufficient to show that conceptually the question of the direction of time must be separated from questions about eternalism. However, it may be that no physical meaning can be attached to the idea of the direction of time in the whole universe, because no global time co-ordinate for the whole universe can be defined. This seems to be implied by special relativity.

The status of time in special relativity (SR) differs from its status in Newtonian mechanics in that there is no objective global distinction between the dimensions of space and that of time. Spacetime can be split into space and time, but any such foliation is only valid relative to a particular inertial frame, which is associated with the Euclidean space and absolute time of the co-ordinate system of an observer. This seems to imply eternalism, since if there is no privileged foliation of spacetime, then there is no global present, and so the claim that future events are not real does not refer to a unique set of events.[8] Furthermore, many have argued that, since SR implies the relativity of simultaneity, whether or not two events are simultaneous is a frame-dependent fact, and therefore there is no such thing as becoming.[9] However, this is too quick. It is possible to advocate a form of becoming that is relative

[8] Of course, one could argue that the very notion of reality must be relativised to observers, but this is to give up on the kind of metaphysics at issue.

[9] The literature on these topics is voluminous but among the most influential papers are Gödel (1949), Putnam (1967), and Stein (1968) and (1989).

to observers or events, so strictly speaking it is only absolute becoming that is ruled out by the lack of absolute time in special relativity. Since the light cone structure of Minkowski spacetime is Lorentz invariant, it can be regarded as absolute. It is easy then to define a notion of the open future of an event E, since any event E' in the forward lightcone of E will have events in its backwards lightcone that are not in the backwards lightcone of E, meaning that there is a sense in which it can be claimed that E' is not determinate at E. This notion of becoming is objective in the sense that all observers will agree about which events are in the open future of a given observer at a particular point in his or her history, because all observers agree about the light cone structure of spacetime.[10]

In any case there is a fundamental problem with drawing metaphysical conclusions about the nature of time from special relativity, namely that it is a partial physical theory that cannot describe the whole universe.[11] For that we must turn, in the first instance, to general relativity (GR), and the implications of that theory for time are not clear. This is because GR gives us field equations that are compatible with a variety of models having different global topological features, and different topological structures may have very different implications for the metaphysics of time. For example, if the topology of the universe is globally hyperbolic then it is possible to define a single global foliation of spacetime for it; otherwise it may not be. Clearly we must then turn to cosmological models of the actual universe, of which there are many compatible with the observational data. As yet there is no agreement about which of these is the true one. Highly controversial issues about the cosmological constant and so-called

---

[10]   Another possibility is to argue that while Special Relativity is empirically adequate, the empirical evidence is nonetheless compatible with the existence of a privileged foliation. This would be to advocate along the lines of that originally proposed by Lorentz and Fitzgerald. This is the strategy adopted by Tooley (2000) for example. Naturalism forbids the revision of scientific theories on purely philosophical grounds, so the proposal of a privileged foliation contra to SR requires a scientific motivation. One possible scientific motivation is the adoption of a solution to the measurement problem that posits a preferred frame of reference. Another is the identification of foundational problems with the account of relativistic phenomena in the Minkowski spacetime framework (see Bell 1987 and Brown 2005).

[11]   Syder (2001) is a recent example of a metaphysician who confines discussion to the implications of Special Relativity.

dark energy, dark matter, and the nature of singularities, as well as the various approaches to the search for a theory of quantum gravity, all bear on the question of whether spacetime will turn out to admit of a global foliation, and hence on whether absolute time is physically definable. Even if it does turn out to be definable, there remains the question of whether such a definition ought to be attributed any metaphysical importance. For example, it is possible to define something called 'cosmic time' which is based on the average properties of the universe's global matter distribution under the expansion of the universe. Some have argued that we can regard Cosmic Time as giving us a privileged foliation (Lucas and Hodgson, 1990). However, others argue that the fact that such a foliation can be defined gives us no reason to regard it as having an objective significance (Berry, 1989), not least because it is based on averages that have nothing to do with the phenomenological experience of everyday simultaneity for objects whose states of motion are not the same as the state of motion of the galaxies in our region of the universe with respect to which cosmic time is defined (see Bourne, 2004).

Non-relativistic many-particle quantum mechanics does not directly bear on the philosophy of time since the status of time in the formalism is not novel in the same way as in relativity. However, it has often been argued that quantum physics is relevant to questions about the openness of the future, becoming, and the direction of time, because of the alleged process of collapse of the wavefunction. Since Heisenberg (1962) it has been popular to claim that the modulus squared of the quantum mechanical amplitudes that are attached to different eigenstates in a superposition represent the probabilities of genuinely chancy outcomes, and that when a measurement is made there is an irreversible transition from potentiality to actuality in which the information about the weights of the unactualised possible outcomes is lost forever. Hence, measurement can be seen as constituting irreversible processes of becoming that induce temporal asymmetry. However, quantum measurements need not be so understood not least because some deny that collapse is a genuine physical process and also because realism about the wavefunction is highly contentious. Similarly, if there is no collapse, as in the Everett interpretation, then again there is no temporal asymmetry in quantum mechanics.

The upshot seems to be that the status of the arrow of time in quantum mechanics is open.

There is also a vast literature about whether or not the second law of thermodynamics represents a deep temporal asymmetry in nature. The entropy of an isolated system always increases in time, and so this seems to be an example of the arrow of time being introduced into physics. If the whole universe is regarded as an isolated object, and if it obeys the second law, then it would seem that there is an objective arrow of time in cosmology. However, it is not clear what the status of the second law is with respect to *fundamental* physics. One possibility is that the second law holds only locally, and that there are other regions of spacetime where entropy is almost always at or very near its maximum. (Boltzman himself thought this was the case in his later years.) Even if thermodynamics seems to support the arrow of time, it is deeply puzzling how this can be compatible with an underlying physics that is time asymmetric. Conservative solutions to this problem ground the asymmetry of the second law in boundary conditions rather than in any revision of the fundamental dynamics. The most popular response is to claim that the law does indeed hold globally but that its so doing is a consequence of underlying time-reversal invariant laws acting on an initial state of the universe that has very low entropy. This is called the 'Past Hypothesis' by Albert (2000). A much more radical possibility is that the second law is a consequence of the fact there is a fundamental asymmetry in time built into the dynamical laws of fundamental physics. Given the outstanding measurement problem in quantum mechanics those who propose radical answers to problems in thermodynamics and cosmology often speculate about links between them and the right way of understanding collapse of the wavefunction. Both the question of eternalism, and the arrow of time, lead inexorably to cosmology and thence to the realm of quantum gravity.

5. Quantum Gravity[12]

A theory of quantum gravity must do all of the following: say what happens in nature at the Planck length (10^{-33}cm); recover GR as a low-energy limit; and provide a background spacetime, at least phenomenologically, for conventional quantum theories. What else it should do is a matter of great contention. Some, such as advocates of string theory and M-theory like Brian Greene (2004), think it must also unify the four fundamental forces of nature;

[12] Introductions to the philosophy of quantum gravity include Callender and Huggett (2001) especially the introduction, and Rovelli (2004), chapter 1. Wallace (2000) is also very helpful.

others, such as Smolin (forthcoming), argue that, in the first instance at least, it need only amount to a quantised version of GR. A further question is whether quantum gravity must also be a cosmological theory of one (unique and actual) universe, rather than allowing for models representing a variety of universes.

Quantum gravity must reconcile a number of profound tensions between GR and quantum theories. Most obviously, quantum physics is the physics that best describes the phenomena when we look at very short length scales, and GR is physics that was specifically designed with a distinction between local and global properties of spacetime in mind, and sought to describe deviations from the topological, geometrical and metrical properties of Minkowski spacetime that only show up in large scale structure— this is the scale tension. Secondly, GR depends on the identification of inertial and gravitational mass, and the equivalence between accelerating and gravitational frames, whereas quantum theory was originally developed to account for the interaction between electro-magnetic radiation and matter. Initially, it was only the energy states of matter that were quantised, but subsequently it has proved possible, with differing degrees of success, to quantise all the fundamental forces, with the exception of gravity—this is the force tension. Thirdly, relativistic theories obey the condition that there are invariant, and hence objective causal pasts for events, whereas in QM there are nonlocal correlations that some regard as evidence of action at a distance—this is the causal tension. Finally, there is the radically different status of time in quantum theory versus GR. In the former, time is a parameter external to all physical systems; in the latter it is a co-ordinate with no particular physical significance. More specifically there is something called 'The Problem of Time' the upshot of which is that theories of quantum gravity are in danger of saying there can be no change in the universe over time (I return to this below).

Depending on who one listens to, string theory has either already led us a considerable distance down the road to a complete theory of quantum gravity, or it has achieved absolutely nothing that counts as physics rather than mathematics. String theorists have followed the methodology that was used in the construction of quantum field theories, namely the search for fundamental symmetries. If the string theory vision is correct then the ultimate fundamental physics will describe the universal symmetries of the universe. (Another important commonality between string theory and classical and quantum physics is that they posit a continuous space and time, which is departed from by some rival programmes—see below.)

Lee Smolin is highly critical of string theory and argues that it is not falsifiable in the sense that it makes no 'falsifiable predictions for doable experiments' (2006, 197). He claims there are no fundamental global symmetries, on grounds that those theories that posit them are not fully empirically adequate. His view is that the two big ideas that drive string theory, namely unification and symmetry, have run their course, and that there have been no substantially new results in particle physics since 1975. He points out that the standard model has so many adjustable parameters that any likely experimental data from particle accelerators can be accommodated by it.[13] He also emphasises that there are at least 10^{100} possible string theories, and argues that all the ones that have been studied disagree with the data. Super-symmetric string theory has 105 free parameters. Hence, Smolin claims, partisans will be able to maintain that whatever comes out of the next generation of particle accelerators confirms super-symmetry.

He also criticises string theory for being 'background dependent' in the sense that it relies on a background spacetime structure. Smolin predicts that the correct theory of quantum gravity will be relational in the sense that it won't posit any background structure which does not change with time but which is necessary for the definition of kinematical quantities and dynamical laws (ibid.). Newtonian mechanics, SR, quantum theories including quantum field theories, and string theory are all background-dependent and rely on various structures that are outside the scope of the dynamics of the theory. For example, in ordinary quantum mechanics the spacetime and the algebraic structure of Hilbert space are part of the background structure. On the other hand, GR, understood as a cosmological theory of the whole universe, is a relational theory in the sense that the physically important structural features of the theory are dynamical. The only background structure in GR consists of the dimensionality, the differential structure, and the topology.

String theorists now seem to have accepted that background independence is a desideratum. Brian Greene speculates about a background free version of string theory, and the search for so-called M-theory is partly motivated by the need for a way of

[13] There are at least 19 free parameters in the standard model plus almost as many from cosmology.

thinking about strings that does not treat them as vibrations in a background spacetime. However, no such theory yet exists. Meanwhile, Smolin has inspired a significant minority of researchers to seek background independence in other approaches to quantum gravity, and he shows how this notion plays out in the context of a variety of these theories. He suggests that the history of physics testifies to the success of the pursuit of background independence. It is true that progress has sometimes been made in physics by eliminating background structure. SR eliminated the background structure of absolute space and time, and then GR eliminated the background structure of Minkowski spacetime. On the other hand, there have been many background dependent theories that have been highly successful, including quantum theories. Smolin himself concedes that GR is background dependent in certain respects. Furthermore, consider the success of the pursuit of symmetry and unification of forces that motivates string theory, in generating the standard model and the unified field theories based on the knitting together of the symmetry groups previously discovered to be governing the separate forces. Smolin and Greene's dispute can be construed as concerning which of the following two desiderata for fundamental physics holds trumps: symmetry or background independence. The empirical evidence is equivocal, to say the least.

Among background independent approaches the most well-known is canonical quantum gravity. This approach seeks a quantum theory of gravity, but not necessarily a unification of all the fundamental forces. This gives rise to the famous Wheeler-DeWitt equation, and the infamous Problem of Time: the physical states of the universe must be time-independent, and so nothing changes. The latest version of the canonical programme is loop quantum gravity. The pioneers of this approach include Abhay Ashtekar, Carlo Rovelli, John Baez, and Lee Smolin. Other approaches include:

Causal set theory: This is a background independent approach motivated by the assumption that at the Planck scale spacetime geometry will be discrete, and by the fact that a discrete causal structure of events is almost sufficient to define a classical General Relativistic spacetime (see Malament 2006). The formalism models spacetime as a partially ordered set of primitive elements with a stochastic causal structure representing the probabilities for 'future' elements to be added to a given element. The 'volume' of spacetime is then recovered from the number of elements. The 'dynamical structure' is compatible with eternalism because the whole of spacetime can be considered as a single mathematical structure, and temporal relations regarded as just the order of elements. The probabilistic structure is required to be local. It can be shown that a classical spacetime can always be approximated by a causal set. However, the converse does not hold and this is a major problem for this approach (see Smolin 2006). Note that a causal set is a partially ordered set where the intersection of the past and future of any pair of events is finite: '[T]he fundamental events have no properties except their mutual causal relations.' (210).

Causal triangulation models: These models use a combinatorial structure of a large number of 4-simplexes (the 4-d version of a tetrahedron), from which a classical spacetime will emerge as a low energy limit, and from which quantum theory can be recovered if background assumptions are made. Interestingly, the latest simulations of spacetime emerging dynamically from these models generate it as four-dimensional on large scales, but two-dimensional at short distances (and it is known that a quantum theory of gravity is renormalisable in two dimensions).

Topological quantum field theories, twister theory, and non-commutative geometry: these approaches are highly abstract and speculative at present.

Another approach worth mentioning has the unique feature that the temporal dimension is abandoned altogether, namely Julian Barbour's relationism: Time is supervenient on change, but change is just differences between distinct instantaneous three dimensional spaces.

All of these research programmes use new and highly abstract mathematical structures to describe the universe, and theorists hope to get the familiar behaviour of spacetime and quantum particles to emerge as limiting behaviour. It seems clear that we cannot yet say what the metaphysical implications of quantum gravity are, but the possibilities range from eleven dimensions to two, from a continuous fundamental structure to a discrete one, and from a world with universal symmetries to one with none.

With respect to the metaphysics of time, it seems that it is an open question whether there is a objective global asymmetry in time, and whether such dynamical structure as there is in the universe reflects a fundamentally tensed reality or whether eternalism is true. (Mathematically, perhaps the real issue is between three—or more—plus one dimensions and four.) If M-theory is the correct theory of quantum gravity then there will be universal symmetries that are not time dependent, and they may define a background independent structure. If there is no background independent structure and asymmetry in time is part of fundamental physics, then it may be that there is 'dynamics all the way down' and reality is fundamentally tensed.

One possible motivation for the dynamical view is a principle that van Fraassen sometimes seems to endorse, namely that there is nothing that's both perfectly general about all of reality and also true. This coheres with the idea that there is dynamics all the way down in the universe, since any fundamental properties that hold generally would necessarily be time-independent and hence amount to background structure. Consider Smolin: 'The universe is made of processes, not things' (2001, 49). Smolin insists that a lesson of both relativity theory and quantum theory is that processes are prior to states. Classical physics seemed to imply the opposite because spacetime could be uniquely broken up into slices of space at a time (states). Relativity theory disrupts this account of spacetime and in QM nothing is ever really still it seems, since particles are always subject to a minimum amount of spreading in space and everything is flux in quantum field theory, within which even the vacuum is the scene of constant fluctuations.

However, one of the lessons of quantum gravity is that some philosophers—especially standard scientific realists—have jumped to overly strong metaphysical conclusions on the basis of not taking account of all the possibilities still held open by physics. There are two leading examples where what some philosophers treated as decisive rulings from physics are now questioned. One of these is the case we have just been discussing, namely, whether or not we live in a block universe. The other is the alleged discovery by quantum theory that the world is not deterministic. In Bohm theory and the Everett approach, the world comes out deterministic after all. Clearly, theories that seem to wear their metaphysical implications on their sleeves often turn out to admit of *physical* reconstruction in different terms. Many physicists have attempted to resolve tensions between QM and GR by seeking what can be regarded as the key metaphysical truth that lies behind each theory's empirical success. For example, Barbour and Smolin think that relationism is the basis for the success of relativity theory. Often it is argued that the truth of quantisation that lies behind the empirical success of QM means that theorists should pursue discrete structures of space and time as in the programmes of causal set theory. If each instant is ontologically discrete then why should the timeline be continuous? On the other hand, the key insight of QM might also be regarded as the superposition principle, and the consequent problem of entangled states.

So it is not clear which aspects of the metaphysical foundations of contemporary physics—for example, continuous space and time, or four-dimensionalism—will be preserved in quantum gravity. Given the diversity of philosophical and foundational presuppositions and implications currently abroad—for example, ranging from a return to absolute time to nihilism about time, and from ten dimensional continuous spaces to discrete graphs—there is little positive by way of implications for metaphysics that we can adduce from cutting edge physics. However, when it comes to negative lessons, there is more to be said. None of the existing contenders for a theory of quantum gravity is consistent with the idea of the world as a spatio-temporal manifold with classical particles interacting locally. This is not surprising to anyone who has thought about the implications of Bell's theorem. It is important that this is not a theorem about quantum mechanics, but rather tells us something about any possible empirically adequate successor to quantum mechanics, namely that it cannot be both local and posit possessed values for all measurable observables. We are justified in treating as unmotivated the idea that any theory of quantum gravity will be a local realist theory, and we should restrict consideration in metaphysics to theories that are compatible with the violation of Humean supervenience implied by entanglement.

6. Scientific Realism and the Metaphysical Content of Physical Theories.

Here is Larry Sklar:

> ... it is a great mistake to read off a metaphysics superficially from the theory's overt appearance, and an even greater mistake to neglect the fact that metaphysical presuppositions have gone into the formulation of the theory in the first place. (Sklar 1981, 131)

Steven French is another philosopher of physics who has expressed scepticism about the attempt to learn metaphysical lessons from physics. French argues that it is not possible to read the answers to metaphysical questions off our best physics because there is an underdetermination of metaphysics by physics (1998).

It turns out that when we examine any of the instances of alleged metaphysical knowledge being delivered by scientific theories, there are always a number of extra assumptions needed to derive the conclusion which we can contest. So, for example, Einstein contested indeterminism in the face of quantum mechanics by denying that the latter is a complete description of reality. In general, there is the problem of scientific theories underdetermining metaphysics. In the case of quantum mechanics, for example, it is widely known following the work of John Bell that no local deterministic hidden variable theory can replace quantum mechanics and still be empirically adequate. However, a nonlocal hidden variables theory can do so, as is the case with Bohm theory. So we see that a metaphysical package may be ruled out or in, but not metaphysical assumptions taken singly.

It is worth remembering that scientific theories have issued metaphysical claims before, only for subsequent developments to have shown the theories in question to be erroneous in certain fundamental respects. So, for example, Fresnel's ether theory is now regarded as featuring a central theoretical term, 'ether', which does not refer to anything in the world. Given this the best we can say about our best theories is that they are approximately true. This may give cause for doubt that we ought to believe their metaphysical implications.

Recent realist responses to the arguments against scientific realism from theory change have separated the parts of theories that essentially contributed to their novel predictive success from the 'idle' parts of theories. The problem for our project of learning metaphysics from physics is that the idle parts often turn out to be the metaphysical parts. For example, Psillos (1996) argues that in the case of caloric theory all the important predictive successes of the theory were independent of the assumption that heat is a material substance, and in the case of the ether, the predictive success of Fresnel's wave theory of light were independent of the assumption that the ether is a material solid. The other problem with the historical record is that the pessimistic meta-induction becomes much stronger if we confine ourselves to considering the fate of metaphysical posits in the history of science rather than theoretical posits in general. Claims about the existence of unobservable entities may not often be renounced but claims about the metaphysical nature often are.

Other problems in reading off our metaphysics from our best science include the following:

(a) There is no fundamental unity in physics since quantum mechanics and general relativity have not so far been conjoined consistently. This might suggest that we should suspend judgement about the metaphysical implications of both theories until we see which are carried over to their successor.

(b) The measurement problem in quantum mechanics means that we lack a consistent interpretation of the theory that is agreed upon by most physicists.

(c) Scientific theories often come in multiple formulations which may suggest different metaphysical pictures, and theories are only applied to the world via models. So we might doubt whether the empirical success of science ought to make us believe that a particular fundamental theory is true.

(d) Metaphysical component of scientific theories makes the problem of defining approximate truth come to the fore.

The topic of this paper was addressed by Sir James Jeans in 1942 in his book *Physics and Philosophy*. Towards the end comes this passage:

> There is a temptation to try to round off our discussion by summarizing the conclusions we have reached. But the plain fact is that there are no conclusions. If we must state a conclusion, it would be that many of the former conclusions of nineteenth-century science on philosophical questions are once again in the melting pot. (216)

He continues:

> ... physics and philosophy are at most a few thousand years old, but probably have lives of thousands of millions of years stretching away in front of them. They are only just beginning to get under way, and we are still, Newton's words, like children playing with pebbles on the sea-shore, while the great ocean of truth rolls, unexplored, beyond our reach. (217)

I can offer no better conclusion than his:

> ... to travel hopefully is better than to arrive. (217)

References

Albert, D. (2000), *Time and Chance*, Cambridge, MA: Harvard University Press.

Berry, M. (1989), *Principles of Cosmology and Gravitation*, Bristol: Adam Hilger.

James Ladyman

Bourne, C. (2004), 'Becoming inflated', *The British Journal for the Philosophy of Science* 55, 107–119.

Callender, C. and Huggett, N. eds. (2001), *Physics Meets Philosophy at the Planck Scale: Contemporary Theories in Quantum Gravity*, Cambridge: Cambridge University Press.

Couvalis, S. (1997), *The Philosophy of Science: Science and Objectivity*. London: Sage.

Dummett, M. (2000), 'Is Time a Continuum of Instants?', *Philosophy* 75, 497–515.

French, S. (1998), 'On the withering away of physical objects', in E. Castellani, ed., *Interpreting Bodies: Classical and Quantum Objects in Modern Physics*, 93–113. Princeton: Princeton University Press.

Friedman, M. (1999), *Reconsidering Logical Positivism*, Cambridge: Cambridge University Press.

Gödel, K. (1949), 'A remark about the relationship between Relativity Theory and idealistic philosophy', in A. Schilpp, ed., *Albert Einstein, Philosopher-Scientist*, 557–562, La Salle: Open Court.

Greene, B. (2004), *The Fabric of the Cosmos: Space, Time and the Texture of Reality*, London: Allen Lane.

Hawley, K. (2006), 'Science as a Guide to Metaphysics', *Synthese*, 149, 451–470.

Heisenberg, W.(1962), *Physics and Philosophy*, New York: Harper and Row.

Jackson, F. (1998), *From Metaphysics to Ethics: A Defence of Conceptual Analysis*, Oxford: Oxford University Press.

Jeans, J. (1943), Physics and Philosophy, Republished by New York: Dover 1981.

Ladyman, J. and Ross, D. (2007), *Every Thing Must Go: Metaphysics Naturalised*, Oxford: Oxford University Press.

Lowe, E.J. (2002), *A Survey of Metaphysics*, Oxford: Oxford University Press.

Malament, D. (2006), 'Classical General Relativity', in J. Butterfield and J. Earman, eds., *Handbook of the Philosophy of Physics*, Elsevier.

Lucas, J. and Hodgson, P. (1990), *Spacetime and Electromagnetism*, Oxford: Oxford University Press.

Maudlin, T. (2002b). Remarks on the passing of time. *Proceedings of the Aristotelian Society* 102: 237–252.

Putnam, H. (1967), 'Time and physical geometry', *Journal of Philosophy* 64, 240–247.

Psillos, S. (1996), *Scientific Realism: How Science Tracks Truth*, London: Routledge.

Reichenbach, H. (1956), *The Direction of Time*. Berkeley: University of California Press.

Rovelli, C. (2004), *Quantum Gravity*. Cambridge: Cambridge University Press.

Sider, T. (2001), *Four-Dimensionalism: An Ontology of Persistence and Time*, Oxford: Oxford University Press.

Sklar, L. (1981), 'Time, Reality and Relativity', in R. Healey (ed.), *Reduction, Time and Reality*, Cambridge: Cambridge University Press.

Smolin, L. (2006), 'The relational idea in physics and cosmology', in D. Rickles, S. French and J. Saatsi, *Structural Foundations of Quantum Gravity*, Oxford: Oxford University Press.

Stein, H. (1968), 'On Einstein-Minkowski space-time', *Journal of Philosophy* 65, 5–23.

Stein, H. (1991), 'On relativity and openness of the future'.

Tooley, M. (2000), *Time, Tense and Causation*. Oxford: Oxford University Press.

Van Fraassen, B.C. (1991), *Quantum Mechanics: An Empiricist View*. Oxford: Oxford University Press.

Van Fraassen, B.C. (2002), *The Empirical Stance*. New Haven: Yale University Press.

Wallace, D. (2000), 'The quantization of gravity—An introduction' unpublished manuscript. available at arXiv:gr-qc/0004005 v1.

Watkins, J. (1975), 'Metaphysics and the Advancement of Science', *The British Journal for the Philosophy of Science*, 26, 91–121.

Natural Kinds: Rosy Dawn, Scholastic Twilight

IAN HACKING

The rosy dawn of my title refers to that optimistic time when the logical concept of a natural kind originated in Victorian England. The scholastic twilight refers to the present state of affairs. I devote more space to dawn than twilight, because one basic problem was there from the start, and by now those origins have been forgotten. Philosophers have learned many things about classification from the tradition of natural kinds. But now it is in disarray and is unlikely to be put back together again. My argument is less founded on objections to the numerous theories now in circulation, than on the sheer proliferation of incompatible views. There no longer exists what Bertrand Russell called 'the doctrine of natural kinds'—one doctrine. Instead we have a slew of distinct analyses directed at unrelated projects.

First thesis:

Some classifications are more natural than others, but *there is no such thing as a natural kind*.

That, in the smallest number of words, is exactly what I mean. Rigour demands more words. In the language of classes, there is no well-defined or definable class whose members are all and only natural kinds. Likewise there is no fuzzy, vague, or only loosely specified class that is useful for any established philosophical or scientific purpose, and which is worth calling the class of natural kinds.

Nelson Goodman was right. If the word 'kind' is to be used as a free-standing noun with a grammar analogous to 'set'—a practice introduced by William Whewell in 1840—there are only relevant kinds.

I say 'relevant' rather than 'natural' for two reasons: first, 'natural' is an inapt term to cover not only biological species but such artificial kinds as musical works, psychological experiments, and types of machinery; and second, 'natural' suggests some absolute categorical or psychological priority, while the kinds in question are rather habitual or traditional or devised for a new purpose.[1]

Goodman wanted philosophers to realize that many questions posed in the context of natural kinds—induction for example— arise equally for other kinds of things, such as machinery or musical works. In consequence he may have overstated the case, and be taken to imply that there is no point in distinguishing some kinds, in a variety of contexts, as natural.

Obviously, what makes a class relevant to a person or a community may be facts about nature in the wild, on the farm, in the stars, in the lab, in the human psyche, or in the nucleus of an atom. Goodman asserts only that we cannot proceed in a general way, beyond the fact that some kinds are relevant for this or that purpose. Some kinds are relevant because of their role in systematic biology. Among the questions that arise: are our classifications natural in the sense that they represent morphology and function, or in the sense that they capture evolutionary history? Some kinds are relevant because of their role in experimental and theoretical physics. Philosophers tend to single out as natural those that are profound, or fundamental, or, in Quine's word, cosmic. Polymer science, cognitive psychology, and silviculture all have their kinds which are relevant to the varied interests of current research and application. Meteorologists, seamen and peasants distinguish cirrus from altostratus from cumulonimbus, kinds of clouds found in nature and even today very useful for predicting the weather. Natural kinds of clouds, we might say, as opposed to those found in a cloud chamber for studying cosmic rays—and yet the cloud chamber was developed for studying—clouds.

Certain artificial crystals are superconducting at (relatively) high temperatures. No one understands why. In one sense these crystals are not a natural kind: they are made in the laboratory. In another sense they are a natural kind, precisely because of this fascinating and probably useful property of being superconducting. They are subject to very intense scientific research right now (a criterion sometimes used to distinguish natural kinds). A more pedestrian example in the same vein is Kevlar®: a kind of material substance so artificial that DuPont holds the patent. It is relevant to policemen and canoeists because if is light and very sturdy,

---

[1] N. Goodman, *Ways of Worldmaking* (Indianapolis: Hackett, 1978), 10.

resisting both bullets and jagged rocks in the white water. It is also relevant to materials science where its properties are still under investigation using tools such as the atomic force microscope for examining the unique structure of its surface.

Goodman's expression, 'habitual or traditional or devised for a purpose' does not do justice to the variety of kinds of relevance to students of nature, but his instinct was right. There is no such thing as the class of natural kinds. This first thesis may be the other side of the coin on which John Dupré inscribes his pluralism, which he formerly called promiscuity.[2] He urges that even in the case of the life sciences there are some and maybe numerous cross-cutting classifications that yield classes worth calling natural kinds. Morphological kinds and evolutionary kinds, to take two examples just mentioned. My argument is that there are so many radically incompatible theories of natural kinds now in circulation that the concept itself has self-destructed. The reverse side of the coin, Dupré's, may be less disconcerting than my obverse. If so, I urge the reader to accept Dupré, and then turn the coin over, and acknowledge that the concept of a natural kind, which began in a promising way and has taught us many things, is now obsolete.

Humpty Dumpty

Stipulative definitions are always open for those who wish to wax rhetorical. Humpty Dumpty can call any class he fancies the class of natural kinds. I, for instance, am at present very keen on bosons and fermions. I am wont to say, not entirely falsely, that everything is either a boson or a fermion, or a species of one or the other. I can express my high regard with an even greater flourish, saying that there are exactly two natural kinds in the universe, *boson* and *fermion*, and, derivatively, their species. Now *that* is a well-defined class! (Yes, physicists do speak of species. Fermions tend to be light, such as electrons, but in this context, an experimenter who speaks of species is likely referring to atoms or ions of an isotope.

[2] J. Dupré, *The Disorder of Things: Metaphysical Foundations of the Disunity of Science* (Cambridge, Mass.: Harvard University Press, 1993). 'In Defence of Classification' *Studies in the History and Philosophy of the Biological and Biomedical Sciences* **32** (2001), 203–219; reprinted in *Humans and Other Animals* (Oxford: Clarendon, 2002), 81–99. 'Is "Natural Kind" a Natural Kind?', *The Monist* **85** (2002), 29–49, reprinted in *ibid.*, 103–123.

Ian Hacking

Every element except Beryllium has isotopes that are bosons. Every element has isotopes that are fermions. So the unexpected talk of species suits: Rubidium 47 is a species of boson.) I have, however, added nothing to our understanding of *anything*, by calling this class of fundamental kinds of entities the class of natural kinds.

Notice that I have not used a wholly arbitrary class as my example. There are good reasons to say that bosons, fermions and their species are curiously fundamental to the universe as (for the moment) we know it.

Philosophical research programmes connected with natural kinds have brought many logical truths to light. They have fallen on hard times. They have split into sects, to the extent that paradigm natural kinds for one set are not natural kinds at all for another. The doctrine of natural kinds is in such disarray that it does tend to humptyism. Advocates will refer to the class of classifications they most admire, as the class of natural kinds. The class is often of great interest. Yet the chief reason for calling it the class of natural kinds is that that sounds good. It confers a rhetorical pedigree on the class. When natural kinds become redefined as some special-interest class, one is tempted to invoke Imre Lakatos's phrase, and speak of a degenerating research programme.

Second thesis:

Many philosophical research programmes have evolved around an idea about natural kinds, but *the seeds of their failure (or degeneration) were built in from the start.*

There are standard examples of natural kinds. Familiar ones from the 1970s include: tigers, lemons, water, gold, multiple sclerosis, atoms, heat and the colour yellow. Each of these classifications, with the possible exception of multiple sclerosis (which might turn out to be several quite distinct diseases), will have a useful role for the foreseeable future. But there is neither a well-defined class, nor any useful vague class, that collects together these heterogeneous examples in the ways that philosophers have hoped for.

Look no further than the paradigms just cited, and wonder, how could there be a class that fruitfully collects together such a wonderful array of interesting kinds of ...—kinds of what, anyway? *Tiger* names a kind of animal, *lemon* a kind of fruit, and also, in a related sense, a kind of tree. *Gold* names an element, and in another sense a metal, a substance that comes in lumps, dust and flakes.

Natural Kinds: Rosy Dawn, Scholastic Twilight

Beyond the animals, vegetables, and minerals, the standard examples do not seem to be kinds of anything. Of what is heat a kind? The sheer heterogeneity of the paradigms for natural kinds invites scepticism. The question should be: why would philosophers ever have imagined that there is one definite, human-independent, class of natural kinds? There have been good reasons, as we shall see, but it is not a concept to be taken for granted.

A handy tag

My first thesis does not imply that the expression 'natural kind' is useless. It may even be crossing from technical philosophy to more common usage. In May 2001 I read in a *New Yorker* book review that: 'Unlike, for instance, a high school, a decade is not what philosophers call a "natural kind": life does not carve itself up spontaneously into ten-year segments.'[3] No philosopher of science has ever called *high school* a natural kind, but it is perfectly clear what the author, Louis Menand, meant. He was reviewing two books about the Seventies. Our fixation on decades or centuries is conventional. It can be a cheerful or ironical way to identify our generation, our friends, and our times. Or other generations, other times. 'The seventies' is nevertheless an artefact of our decennial dating system. The high school, on the other hand, is a cardinal institution in American life, and for many it marks the time when one grows up, a very natural kind, which has become one of the seven ages of middle-class American men and women.

For another example, in a more philosophical vein, we can go back to Richard Rorty.[4] In an ironic critique of Bernard Williams's position that science and morality are fundamentally different in character, he challenged the thought that natural science is a natural kind. He meant that there is a sort of continuum between moral and scientific reasoning. Both Menand and Rorty made good use of the phrase 'natural kind'. Neither implied that there is a distinct class of natural kinds, only that in this or that context, some kinds are more—or less—deeply rooted than others.

[3] L. Menand [book review], *The New Yorker*, 28th May 2001, 128.
[4] R. Rorty, 'Is Natural Science a Natural kind?', in E. McMullen (ed.), *Construction and Constraint: The Shaping of Scientific Rationality* (Notre Dame, Ind.: Notre Dame University Press, 1988), 49–74. Reprinted in Rorty, *Objectivity, Relativism, and Truth* (Cambridge: Cambridge University Press 1991), 46–62.

Ian Hacking

Finally, in a spirit of what sounds like post-modern reflexivity, John Dupré asks, 'Is "natural kind" a natural kind?'[5] Merely modern persons like me will answer no-or-meaningless, because we tend to practice, even if we do not affirm, the theory of types. For the nonce, let us follow Quine's famous paper and regard natural kinds as sets: 'Kinds can be seen as sets, determined by their members. It is just that not all sets are kinds.'[6] If so, Dupré's question violates any informal theory of types: 'natural kind' cannot apply to itself. That is not to say his discussion fails to cohere. One of his minor conclusions comes to this: if an essentialist wishes to extend his essentialism to second-order talk of natural kinds, then he will answer: 'Yes, the class of natural kinds is a natural kind.' Pluralists will answer: 'No, but nevertheless there are a number of ways in which our interests in natural phenomena lead us to single out some systems of classification as peculiarly natural.'

Perhaps these usages point the way for a modest future role for the expression 'natural kind'. It will no longer be used as an absolute classification that divides sets and their ilk into those that are natural kinds and those that are not. It will no longer have anything special to do with 'Nature', that wondrous world independent of the human mind. It will be used for those classifications that in context strike us as natural, as opposed to those that strike us as conventional. A future *OED* could use my quotation from *The New Yorker* for one of its examples.

William James: seven-league boots

I have nothing against classification. I love the richness of the different kinds of things there are in the world and of the innumerable ways in which they can be grouped together. Fools and poets may see this best, as chanted in Nietzsche's wild paean, *Only a fool! Only a poet!* in praise of, among other things, the fierceness

[5] 'Is "Natural Kind" a Natural Kind?', *op. cit.* note 2.

[6] W.V.O. Quine, 'Natural Kinds', *Ontological Relativity and Other Essays* (New York: Columbia University Press, 1969), 114–138, on 118. Note that after the first few pages, Quine drops the adjective 'natural' and writes of kind and kinds. One could propose this as the truly pragmatist way of speaking, a return to William James, as discussed in the next section.

of variety.[7] It can be read as an apocalyptic version of the gentle wisdom of Hopkins's equally philosophical *Pied Beauty*. Protestant children still sing, 'All things bright and beautiful, / All creatures great and small', another hymn to variety that is not be despised. But best of all, for we more prosaic thinkers, is the good old Yankee common sense of William James:

> Kinds, and sameness of kind—what colossally useful *denkmittel* for finding our way among the many! The manyness might conceivably have been absolute. Experiences might have all been singulars, no one of them occurring twice. In such a world logic would have had no application; for kind and sameness of kind are logic's only instruments. Once we know that whatever is of a kind is also of that kind's kind, we can travel through the universe as if with seven-league boots.[8]

James spoke of kinds, not natural kinds. It is the idea of a well-defined class of natural kinds that has self-destructed, not kinds and sameness of kind.

Dawn

Natural History: species become absolute

The doctrine of natural kinds grew out of the problem of natural groups. That had two origins, travel, and an intellectual innovation. Exploration brought to European shores ever so many new kinds of plants, animals and rocks. To quote James from another context, 'a blooming, buzzing confusion'. The sheer proliferation of fauna and flora and minerals created a demand for classificatory systems.

---

[7] *Nur Narr! Nur Dichter!* in F. Nietzsche, *Dithyrambs of Dionysus*, bilingual edition, translated by R. J. Hollingdale from *Dionysos-Dithyramben* (1891), (London: Anvil Press, 1981, 22–27). Hollingdale translates *bunt* as gaudily or gaudy, as in *ein Tier, ein listiges, raubendes, schleichendes, / das lügen muβ, / das wissentlich lügen muβ, / nach Beute lüstern, / bunt verlarvt*—'lusting for prey, gaudily masked'. Or *Nur buntes redend*—Talking only gaudy nonsense. I do not wish to argue with such a masterly translator, but would add that *bunt* above all is associated with varied bright colours, our bunting on festive sailboats. Joseph's cloak of many colours is *bunt*. It is the *bright variety* that I take from Nietzsche's *bunt* more than the gaudiness.

[8] W. James, Pragmatism: A New Name for Some Old Ways of Thinking (New York: Longmans Green 1907), 179.

More radical historians of systems of thought propose that a felt need to represent the order of things happened to precede the sudden influx. Be that as it may, the intellectual innovation is what matters to us.

In all scholastic classification, going back to Aristotle and beyond, genera and species were relative terms. Porphyry's *Introduction* to Aristotle's *Categories* explained for generations upon generations of schoolchildren that *animal* was a species of *living thing*, but a genus of *rational animal*. *Eidos* and *genos* have a far wider application in Aristotle than logic, but I accept Pierre Pellegrin's definitive analysis according to which Aristotle's biology did not aim at a hierarchy of ranks, or a taxonomy.[9] It was only in the time of Linnaeus that species and genera were made part of an absolute hierarchy. Species fell below genera, and no species could be a genus of something else. The hierarchical structure became irreversible, and with it, higher ranks: families, orders, and classes.

That is when a series of problems arose. Imagine the naturalist sorting specimens into groups, to be called species. There arises a first, descriptive, question: does this sorting represent the way the individuals resemble each other, or does an alternative represent nature better?

Resemblance requires a further term, resemblance in what respect? Many great debates reduce to that. Morphology and function of organs were major candidates for the basic ways in which living things should be classified. Linnaeus's brilliant decision is with us still: we start classifying using resemblance in sexual organs as the key.

The natural historian arranges the groups called species into higher groups, called genera. After that, a whole hierarchy is created, genera grouped into families, families into classes, classes into orders and so on. There arises a second, more ontological, question: are such structures mere conveniences, mnemonics so that naturalists can remember what goes where in the ever expanding tables of representation of an ever expanding collection

---

[9] P. Pellegrin, *Aristotle's Classification of Animals: Biology and the Conceptual Classification of the Aristotelian Corpus*, translated from the French of 1982 by A. Preuss (Berkeley and Los Angeles: University of California Press, 1986). A subtle exposition of the earlier view, that Aristotle was groping for a taxonomy, is found in G. E. R. Lloyd, 'The Development of Aristotle's Theory of the Classification of Animals', *Phronesis* **6** (1961), 59–80.

of specimens? Or is there a right arrangement, which shows how nature truly is? Are genera real? Buffon wrote in 1749, that nature

> knows none of these pretended families, and contains in fact nothing but individuals.[10]

Linnaeus, in 1751:

> Revelation, observation and thought confirm that all genera and species are natural. All genera are natural, and have been such since the beginning of time.[11]

Readers today note the assumed (and wrong) eternal fixedness of species, but in mid-eighteenth century, the debate was whether the arrangement of individuals into species and genera and families told the truth about nature. Michel Adanson, in 1763:

> I do not know how any botanist can maintain such a thesis: it is certain that until now no one has been able to prove it, or to give a definition of a natural genus, but only of an artificial one.[12]

There you have it. Which groups are natural, which artificial? When naturalists spoke of artificial classes, they did not have in mind the wholly arbitrary classes often mentioned by modern philosophers, to contrast with natural kinds. 'Natural kinds are standardly distinguished from arbitrary groups of objects, such as what you had for breakfast.'[13] Classes were artificial rather than natural, when they had been invented by botanists, but did not accurately represent the order of living things.

Even a smattering of history informs us that ever so many questions were in play. Linnaeus ruthlessly decided that what matters is sex, and used methods of reproduction as the basic tool both for sorting specimens into species, and for sorting species into genera. Adanson said the choice of sex was artificial, and proposed 65 different characters in terms of which plants could resemble each other. The more the characters, by which individuals closely

---

[10] G.-L. L. Buffon, Histoire naturelle générale et particulière, avec la description du cabinet du roi, Vol. IV, Histoire générale des animaux (Paris: Imprimerie Royale, 1753), 355b.
[11] C. Linnaeus, *Philosophia Botanica* (Stockholm: G. Kiesewetter, 1751), 100.
[12] M. Adanson, *Histoire naturelle du Sénégal* (Paris: C.-J.-B. Bauche, 1763), xv.
[13] C. Daly, 'Natural Kinds', *Routledge Encyclopedia of Philosophy* CD-ROM (London: Routledge, Version 1.0.)

resembled each other, the more natural was the grouping. Hence he is regarded as the precursor of numerical taxonomy.

I have just distinguished two questions, which I called descriptive and ontological. Once the idea of a taxonomic hierarchy is established, we have ranks—species, genera, families, orders. Suppose that in the botanical or zoological garden, and in the tables that correspond to it, we find an interesting group of groups. That poses a third question, what is its rank? Is it a genus or a family? According to the doyen of the history of systematics, Peter Stevens, these three questions were seldom distinguished.[14] Collectively one asked: *what is a natural group?*

The polemics continued, with no definitive resolution, until 1859, when Darwin asked:

> Naturalists try to arrange the species, genera, and families in each class, on what is called the Natural System. But what is meant by that system?

We all know his answer:

> All true classification is genealogical; that community of descent is the hidden bond which naturalists have been unconsciously seeking. [...] The *arrangement* of the groups within each class, in due subordination and relation to the other groups, must be strictly genealogical in order to be natural.[15]

This answer leaves wide open what the true genealogical order is. There are just as many polemics about correct and incorrect phylogenetic trees today as there were about natural classification two centuries ago, maybe more. But aside from new and vexing issues about lateral gene transfer among the earlier organisms, such as bacteria, we are agreed on what makes a group natural. In 1859 the species ceased to be objects that represent nature in terms of resemblance, and became historical objects. We might have to replace the post-Darwinian structure of a tree of life by the model of an estuary, but the objects will still be historical.[16]

---

[14] P. Stevens, *The Development of Biological Systematics: Antoine-Laurent de Jussieu, Nature, and the Natural System* (New York: Columbia University Press, 1994), 10–13.

[15] C. Darwin, *The Origin of Species* (London: John Murray 1859), 413, 420.

[16] For a popular account of post-tree architecture, see W. F. Doolittle, 'Uprooting the Tree of Life', *Scientific American*, February 2000, 90–95. For the estuary model, and other reasons for not sticking with trees, see

Natural Kinds: Rosy Dawn, Scholastic Twilight

Miscellaneous hierarchies

One more word about species before we move to natural kinds. The hierarchical model was not restricted to plants and animals. Using exactly those words, species and genus, and to some extent the higher taxa, it applied across the board in descriptive sciences. All tried to emulate life. Rocks were for long classified in the same way as plants. The vast halls of mineralogical specimens in the École des Mines in Paris were modelled on the Jardin des Plantes a mile away. There was a resolute desire to organize things into a hierarchy, with each successive room containing a group of subgroups. It did not work. Specimens are now arranged using two non-meshing principles, namely the chemical substances they contain, and their crystalloid shape. The degenerating project of hierarchical classification continued well into the nineteenth century. William Whewell was the man who reintroduced the word 'kind' into logic. His first job? Professor of mineralogy at Cambridge University. In those days it was still plausible to imagine that plants and rocks could be sorted according to the same principles.

The garden of species was planted deep.[17] Take diseases. Medical nosology was patterned on plants. Great nosologists such as Boissier de Sauvages in France and Robert James in England followed Linnaeus—with whom Sauvages corresponded extensively. The Linnaean model continued longest in the classification of mental illness, through Kraepelin to the latest *Diagnostic and Statistical Manual of Mental Disorders*, although as with any other degenerating programme, the *Manual* resorts to more and more epicycles.

Neither last nor least, as the chemists formed the idea of what we call the chemical elements, they tried to sort these into a taxonomy. For example, in 1815 Ampère made a valiant attempt at a 'natural classification of the simple bodies', *les corps simples* being the then

---

Dupré, *Humans and Other Animals*, (*op. cit.* note 2), p. 86. For the origins of tree diagrams in Western logic and science, see Ian Hacking, 'Trees of Logic, Trees of Porphyry', in *Advancements of Learning: Essays in Honour of Paolo Rossi*, J. Heilbron (ed) (Florence: Olschki, 2007), 157–206.

[17] This insightful phrase comes from Michel Foucault, in the title of Chapter I, Part II of *Folie et Déraison: Histoire de la folie à l'age classique*, Paris: Plon, 1961. Finally translated as *History of Madness* (London: Routledge, 2006).

current name for the elements.[18] The result looks just like Linnaeus. Repeated attempts to build tables did in the end succeed. Mendeleev made the radical break with botany. The periodic table is the permanent refutation of the idea that natural kinds have to be organized into a tree-like hierarchy. There are obvious genera and species within the table, for example the halogens form a genus of which chlorine and iodine are species. But the structure is not a simply hierarchic set of nested sets. The tragedy of the final contributions of Thomas Kuhn is that he thought that a theory of tree-like kinds, arranged as genus and species, would explain incommensurability and much else.[19] Unfortunately kinds are not tree-like unless nature makes them so. It has been repeatedly argued that natural kinds must, as a matter of logic, be arranged in a tree-like hierarchy.[20] Not so. Bosons, isotopes, and elements are commonly regarded as natural kinds. But since rubidium-47 is a species both of boson and of rubidium, but rubidium is not a species of boson or vice versa, you cannot put these on a branching tree. The fundamental point is that, as Darwin saw, genealogy does the trick for kinds of living things (or does so until lateral gene transfer kicks in). But little in nature is genealogical except life itself.

Darwin and Mendeleev between them demolished the conceptual structure to which the doctrine of natural kinds emerged as a plausible response.

William Whewell 1840

The question of natural groups was everybody's problem in 1840: botanists, zoologists, mineralogists, and students of disease, of language, of human societies, of races and of the chemical elements. Up stepped a formidable duo to translate the whole thing into philosophy, namely William Whewell and John Stuart Mill.

---

[18] A.-M. Ampère, 'D'une classification naturelle pour les corps simples', *Annales de Chimie et Physique* (n.s.) **1** (1815), 295–309, 373–395, **2** (1816), 105–116.

[19] T. S. Kuhn, 'Afterwords', in P. Horwich (ed.), *World Changes: Thomas Kuhn and the Nature of Science*, (Cambridge, Mass.: MIT, 1993), 311–341. Reprinted in T. S. Kuhn, *The Road since Structure,* J. Conant and J. Haugeland (eds.) (Chicago: University of Chicago Press, 2000), 224–252.

[20] For example, by R. Thomason, 'Species, Determinates and Natural Kinds', *Nous* **3** (1969), 95–101.

Formidable partly because they agreed about almost nothing except kinds. Between them they set the engine of natural kinds in motion. Whewell provoked, but is innocent of, the notion that there is a privileged, well-defined, class of kinds, independent of human interests, that should uniquely be called the natural kinds. That is Mill's contribution, but he probably took his building blocks from Whewell.

The word 'kind' had long disappeared from logic, and had little use in any context as a free-standing noun.[21] Whewell revived it. He was the great word-minter of English science, giving us even the name of the occupation, 'scientist'. Quite a number of his hundreds of more specialised nouns 'took', and are in use today. Whewell thought he could cut across the polemics about natural groups by insisting on logical clarity. Behind all the furore, quite aside from the specific science, we had to ask for the fundamentals of classification. The word he chose for this purpose was 'kind'. With the hindsight of our modern convictions he looks wonderfully prescient. What are kinds? '[...] such classes as are indicated by common names.'[22]

Items are grouped under common names when they are like each other: 'The idea of likeness is perpetually operating to distribute [our sensations] into kinds, at least as far as the use of language requires'. But unlike so many philosophers before and after him, resemblance was not the end of the matter but a question to pose. 'Upon what principle, under what conditions, is the idea of likeness thus operative? What are the limits of the classes thus formed? Where does that similarity end, which induces and entitles us to call a thing a *tree*?'

Whewell did not proceed to say that *tree* must be defined by some set of necessary and sufficient conditions, or that there was a set of properties (other than, trivially, being a tree) that nicely determines which things are trees. Explicit definitions are not usually possible.

---

[21] C. S. Peirce cites Wilson's *Rule of Reason* (1551) and Blundeville's *Arte of Logicke* (1599) for stand alone 'kind' in logic. 'Kind', *Baldwin's Dictionary of Philosophy and Psychology*, (New York: Macmillan, 1903), Vol. I, 600. He might have mentioned Locke, who takes 'kind' to be English for genus, and 'sort' to be English for species; *Essay*, III.i.6.

[22] W. Whewell, *The Philosophy of the Inductive Sciences, Founded upon Their History* (London: Parker, 1847), I, 469. All quoted sentences are found in the first edition of 1840, but I cite the second because it is widely available while the 1840 edition is rare. Book VIII, ch. I, §5, is headed *Kinds*, 469. The third and fourth quotations below are from pages 475 and 471 respectively.

His next section is headed, '*Not made by definitions*'. I commend his discussion to you, but here I shall leave you only with his aphorism: '[...] any one can make true assertions about dogs, but who can define a dog?'

Kinds are classes for which we have names. What determines the application of common names? '[...] the Condition which regulates the use of language is that it shall be capable of being used;– that is, that general assertions shall be possible.' Or again, 'The principle, that *the condition of the use of terms is the possibility of general, intelligible, consistent assertions*, is true in the most complete and extensive sense.'

I summarize Whewell thus: A kind is a class denoted by a common name about which there is the possibility of general, intelligible and consistent, and probably true assertions. By 'general' he need not mean only universal assertions, but also assertions that hold, as Aristotle put it, 'for the most part'. Philosophers will find offensive Whewell's explanation of kinds as denoted by common names. Surely there are many kinds for which we have no names in English or other languages! Yet almost exactly Whewell's definition has become the current definition of 'category' in cognitive science and developmental psychology. In a much reprinted and now classic article we read that a category is picked out by 'lexical entry' (viz. a common name) and that a 'A category is a partitioning or class to which some assertion or set of assertions might apply'.[23]

Whewell was able to stand outside sectarian disputes between the Natural Method and the Artificial System of Classification—labels much in use at the time. What makes a good classification? It should deliver us with kinds, and kinds are classes about which we can frame a number of general, intelligible assertions. That gives a positive (positivist) but wholly realist answer to the underlying issue of natural groups. He goes on to produce specific answers to the descriptive, ontological, and hierarchical question about natural groups, anticipating, among other things, Eleanor Rosch's theory of prototypes.[24]

---

[23] D. L. Medin, 'Concepts and Conceptual Structure', *American Psychologist* **44** (1989): 1469–1481, on 1469.

[24] E. Rosch, 'Natural Categories', *Cognitive Psychology* **4** (1973), 328–350. See, Whewell, 494, heading for § 10: 'Natural Groups given by Type, not by Definition'.

Natural Kinds: Rosy Dawn, Scholastic Twilight

John Stuart Mill 1843

People had been talking about natural groups for a century before Whewell, and he continued to do so. The word 'kind' was his innovation, but he never spoke of natural kinds. We know from the *Autobiography* that as early as 1832 Mill wrote a draft manuscript that

> became the basis of that part of the subsequent Treatise [*A System of Logic*]; except that it did not contain the Theory of Kinds, which was a later addition, suggested by otherwise inextricable difficulties which met me in my first attempt to work out the subject of some of the concluding chapters of the Third Book.[25]

Book III of the *System of Logic*[26] is about induction, but Mill did not propose the Theory of Kinds to 'solve' the problem of induction. He began to insert the material on kinds in 1838, two years before Whewell publicly turned free-standing 'kind' into a technical term of logic. Mill had begun to 'recognize Kinds as realities in nature'.[27] But mere 'kind' would not do. It had to be *real* kinds. In case anyone might miss the point, he went through the last draft of the book writing in a capital 'K': real Kinds. This is the beginning of the cavalcade of superlatives. We start with *real*, pass to *natural*, and on to *genuine, aristocratic, strong, elite, pure*, all of which occur in the literature. It is as if you just can't insist enough that you want nothing but the very best: the highest quality kinds on offer in the universe. As if our problems in defining the class of important kinds stems from hitherto shoddy workmanship.

Mill's Theory of Kinds is embedded in his account of names and classes, but we do not need that to get his immediate thought. Not all useful classes are equal, for, 'we find a very remarkable diversity

[25] J. S. Mill, *Autobiography* (London: Longman, 1873), 191.

[26] J. S. Mill, *A System of Logic, Ratiocinative and Inductive. Being a Connected View of the Principles of Evidence and the Methods of Scientific Investigation* (London: Longman, 1st edition 1843). All 8 editions are collated and printed in Vols. VII and VIII of J. Robson (ed.), *Collected Works of John Stuart Mill* (Toronto: University of Toronto Press, 28 vols. 1965–83). References will be given as '*Logic*', followed by Mill's book, chapter and section number, followed by the page number in the Robson edition. Pagination of Vol. VIII continues that of Vol. VII.

[27] 1838 is the date furnished by Robson, 'Textual Introduction', *Logic*, lxv. It is not altogether clear that Mill used the actual word 'kind' before Whewell published.

in this respect between some classes and others.' A class of one type is picked out by a general name, but its members share at most a few other properties that are not implied by that general name. Call them finite kinds. White cells, white roses, and white paper, what have they in common but that they are white? Not much: 'white things, for example, are not distinguished by any common properties except whiteness; or if they are, it is only by such as are in some way dependent on, or connected with, whiteness.'[28]

Now consider the classes, *animal*, *plant*, *horse*, *phosphorus*, and *sulphur*. Phosphorus differs in innumerable ways from non-phosphorus: 'a hundred generations have not exhausted the common properties of animals or of plants, of sulphur or phosphorus; nor do we suppose them to be exhaustible, but proceed to new observations and experiments, in the full confidence of discovering new properties which were by no means implied in those we previously knew.' Such a class is a *real Kind*. Finite kinds and real Kinds, 'are parted off from one another by an unfathomable chasm, instead of a mere ordinary ditch with a visible bottom'.[29]

Why was Mill so excited by this distinction that he made one class into capital K real Kinds? One reason was that he came to believe, after a lot of scepticism, that the scholastic notion of species and genera makes sense. He did not mean the Linnaean concepts then current in natural history, or what was coming to be called biology, but rather the logical conceptions of the schoolmen. He thought he could naturalize them. The era of Whewell and Mill—as befits the closing years of the industrial revolution in Britain, and what Thomas Kuhn called the second scientific revolution—was an era of naturalization in British philosophy, be it empiricist (Mill) or neo-Kantian (Whewell). Rather than logic determining the species and genera, science would settle which classes are real Kinds. Book I, Chapter vii, §4 of the *System of Logic* is headed: *Kinds have a real existence in nature*. This is the clarion call for naturalism, to which the tradition has been faithful ever since. Science determines which kinds are real or natural, not metaphysics or logic.

There was a second reason for Mill's enthusiasm that I shall mention but not explain. Mill thought that he had naturalized scholastic species and genera. But he remained a notorious nominalist, and believed that his view put a final end to essence. He

---

[28] *Logic*. These two quotations and the next are from I. vii. §4, 122.
[29] *Logic*, 123, with a clause inserted in the 4th edition, revised, of 1856.

thought that the 'immortal Third Book' of Locke's *Essay* had almost killed off essence but not quite. 'A fundamental error is seldom expelled from philosophy by a single victory. It retreats slowly, defends every inch of the ground, and retains a footing in some remote fastness after it has been driven from the open country.'[30] Ever since Kripke's bombshell, many readers have associated the doctrine of natural kinds with some notion of essence. Hence it is often forgotten that it was born as a ruthlessly anti-essential philosophy, and remained so through generations of resolute nominalists such as Venn, Russell, Broad, and Quine.

There was a third reason. Real Kinds, as Mill wrote in the passage quoted from the *Autobiography*, were a lifeline to get him out of a hole in Book III, *Of Induction*. Black crows posed a problem for Mill that is specific to his philosophy, and so his discussion seems strange to us today. The problem about crows has nothing to do with Hempel's paradox of the ravens, even if the example may have been suggested to Hempel by Mill.[31]

Mill's analysis of inductive inference relied on his four methods, which in turn relied on causality. Non-causal propositions of coexistence present a problem. Uniformity in nature is commonly the result of causation. Causes precede effects. But some uniformities are simultaneous. How then can they be due to causation? In some cases they are the joint effects of a common cause—the simultaneous high tides on opposite sides of the earth. When uniformities of coexistence are derived from common causes, their degree of certainty or probability is that of empirical laws.[32]

The trouble is that we have good grounds for believing many universal propositions of coexistence, when we are entirely ignorant of a common cause. Hence we cannot use the four methods to investigate or ground our belief, or to explain why propositions of uniformity are (when not true by definition) merely probable. Mill needs to show that uniformities of coexistence, not known to derive from common causes, likewise have the status of

[30] *Logic*, I. vi. §3, 114; I have used the version in the first three editions rather than the slightly rewritten one of 1856 and thereafter. The reference to Locke is from 115.

[31] Hempel was working on confirmation on his arrival in the United States in 1941, publishing his first essay in *Mind* in 1945, followed by an essay specifically on the paradoxes in 1946. It may be relevant that the German translation of the *Logic* translated 'crow' by *Rabe*, rather than *Krähe*.

[32] 'When uniformities of coexistence are derivative, their evidence is that of empirical laws.' *Logic* III. xxii., heading of §6.

empirical laws. They do not have some superior, logical, or essential certainty. Above all, they do not reflect essential properties! This was of paramount importance for Mill.

> The notion that truths external to the mind may be known by intuition or consciousness, independently of observation and experience, is, I am persuaded, in these times, the great intellectual support of false doctrines and bad institutions. [...] There never was such an instrument devised for consecrating all deep-seated prejudices.[33]

He was here expressing his contempt for the Tory attitude, which would preserve the status quo on the grounds of an intuited sense of what is right and 'natural'. He trashed the arguments from mathematics, and in particular the a priorism of Whewell, early in his *Logic*. But now it could come in the back door, by way of uniformities of coexistence not known to be grounded on common causes.

A real Kind is precisely a class of items in which many independent properties coexist. If they coexist thanks to a common cause, no problem. But suppose there is no common cause? In my opinion the rest of the argument is weak. It does not excite us as it did Mill, but remember, for him the point was ideological as well as logical. Here is how it goes. There is never a difficulty in realizing that a uniformity of coexistence is merely contingent and might be false. Suppose that after all not all crows are black—not thanks to a few sports, albino crows, but to the discovery of crows with red shoulders and yellow tail feathers who inhabit remotest Yorkshire. That is only to suppose that, 'a peculiar Kind, not hitherto discovered, should exist in nature'—which 'is a supposition so often realized, that it cannot be considered at all improbable'.[34] Mill takes pleasure in reminding his readers of the strange kinds of animals and birds being reported from Australia, on an almost weekly basis, by every passing ship.

This part of the *Logic* is a rearguard action against essentialism, and the idea that essential coexistence might have an especially high degree of certainty, perhaps even some sort of necessity. That is one of the forgotten origins of the theory of real Kinds, and hence of the doctrine of natural kinds.

---

[33] *Autobiography*, 225–6.
[34] *Logic*, III. xxii. §7, 585–6.

Natural Kinds: Rosy Dawn, Scholastic Twilight

John Venn 1866, 1889

John Venn, known to many for his diagrams, and to a few because of his understanding of frequencies, was probably the man who, in 1866, turned 'real Kinds' into 'natural kinds'. He was an early advocate of a frequency interpretation of probability, and it was there that he first invoked natural kinds. 'There are classes of objects, each class containing a multitude of individuals more or less resembling one another [...]. The uniformity that we may trace in the [statistical] results is owing, much more than is often suspected, to this arrangement of things into natural kinds, each kind containing a large number of individuals.'[35] Only in the next edition of the book, 1876, did he extend the thought from statistical regularity to 'such regularity as we trace in nature'.

Otherwise he made no use of the notion of a natural kind, and in due course he pretty well demolished it. He often took Mill's *Logic* as his guide, but his own logic textbook of 1889 savaged Mill, both on uniformities of coexistence and on the whole idea of a natural kind. He rightly separated what Whewell and Mill had thrown together as kinds, namely 'natural substances' like gold, and 'natural species or classes such as we find in Zoology or Botany'. In the former case, substances, there are indeed coexistent properties. Although (in 1889) we do not know much about the underlying structure that produces the properties of gold, its colour, degree of smoothness and toughness, ductility and malleability, they are, Venn is sure, 'results of the way in which the molecules are packed together'.

> From the practical point of view, such an analysis as [Mill's, in terms of coexistence] is needless. We are quite ready to admit [...] every natural substance contains a group of coexistent attributes. The practical difficulty does not consist in objectifying them [...] it shows itself rather when we attempt to say what belongs to one of these attributes and what belongs to another, in other words to draw the boundaries between them.[36]

---

[35] J. Venn, *The Logic of Chance: An Essay on the Foundations and Province of the Theory of Probability, with Especial Reference to its Application to Moral and Social Science* (London: Macmillan, 1866), 244. Note that the Second edition, much revised, of 1876, and the Third edition, revised, of 1888 contain the same discussion of natural kinds, but the arrangement of matter in the successive editions is very different.

[36] J. Venn, *The Principles of Empirical or Inductive Logic* (London: Macmillan, 1889), 82.

Ian Hacking

This is pretty damning. Mill's theory demands a sense of the way in which the properties of a real Kind are independent of each other. They are not to be in 'some way dependent on, or connected with' the others. The practical difficulty is how to explain this. C. S. Peirce made this point more sharply:

> Mill says that if the common properties of a class thus follow [as a consequence under a law of nature] from a small number of primary characters which, as the phrase is, *account for* all the rest,' it is not a real kind. [Mill] does not remark, that the man of science is bent upon ultimately thus accounting for each and every property that he studies.'[37]

In short, Peirce judged that the methodological assumption of scientific research is that there are no Millian real Kinds.

Venn turned from substances to plants and animals. He may have been the first author to insist that it is absurd to produce one category, 'natural kind', which spans such diverse items as substances and species. What connects the properties of substances is different from what connects the properties of any kind of living thing. Speaking of 'the colour, the smell, the taste of the peach: the speed, the size, the note of the swallow', he writes that:

> Mill, as we all know, writing in præ-Darwinian days, greatly overrated the distinctness and the ultimate or primitive character of these various attributes. He introduced the technical term of 'natural kinds' to express such classes as these, and those considered above [the substances], putting them on much the same footing in respect of natural distinctness and permanence [...].[38]

He noted scathingly (in the last ellipsis) that this implicit 'doctrine of the fixity of species' was the one point on which Whewell agreed with Mill—if Whewell agrees with you, the empiricist Venn implies, you know you must be wrong. If we took Mill (and Whewell) seriously, 'all the aggregate of successive living beings which constituted one of these natural kinds might be put upon

[37] Peirce, *op. cit.* note 21. Pierce presumably knew Venn on natural kinds for he referred often enough to his *Empirical Logic*. He does not seem to have reviewed that book, as he did the 1866 *Logic of Chance*: 'Here is a book which should be read by every thinking man.' *The North American Review* **105** (July 1867), 317–321, on 317. *Collected Papers of Charles Sanders Peirce*, vol. VIII, 3.

[38] Venn, *Empirical Logic*, 83.

much the same footing as the various specimens of the same mineral which exist upon earth'. *Which is absurd.* Later in the book, he rarely used the term 'natural kind' and when he did, he put it in quotation marks—what we now call 'scare-quotes' or 'shudder quotes'—to indicate disapproval.

Heterogeneity

Mill imagined that there is one class that covers two entirely different types of things, what Venn called 'natural substances' such as gold, phosphorus or sulphur, and 'natural species of classes such as we find in Zoology or Botany'. Mill 'writing in præ-Darwinian days', had an excuse. The excuse had vanished in Venn's day. But even Mill ought to have suspected that something was wrong. Mass nouns *are* different from count nouns, even to those who lack the terminology. Mill had another excuse: he wrote not only in pre-Darwinian but also in pre-set-theoretic days. He could use a notion of class in which *phosphorus* and *horse* are not sharply distinguished in point of logic. Those who follow Quine and think of kinds as sets can say that the kind *horse* has members, individual horses, but what are the members of *phosphorus*?

What set is phosphorus? Some people waffle, as if it did not matter much whether it was lumps of phosphorus, or phosphorus atoms. Or ions? Atoms and ions have the advantage that we can count them—and do, when they are trapped, where we may have six ions in a trap or about 10,000 atoms of the same isotope in a trapped cloud. But if we think of phosphorous as a substance, then it cannot be counted.

Quine, never waffling, valiantly tried to 'regiment' mass nouns in a way worthy of a great scholastic logician. In my opinion, once one goes into detail, the resulting regimented discourse makes little sense of what chemists say, or of what condensed matter physicists say. And as for polymer science! But let us stick with phosphorus. It has at least ten allotropic forms, and probably more as yet unknown. The known ones sort into three groups, white, red, and black. The α-white form has a cubic crystal form, while β-white has a hexagonal structure. Is each of these allotropes a natural kind? When Mill came to think about this, he concluded that 'the allotropic forms of what is chemically the same substance are so many different Kinds; and such, in the sense in which the word

Ian Hacking

Kind is used in this treatise, they really are.'[39] Each allotrope has a lot of properties that arise from its specific structure, for example, exposure to sunlight or heat changes white phosphorus to red, which neither phosphoresces nor ignites spontaneously in air.

If you hold that 'phosphorus' names atoms, then it looks like a count noun. But you cannot do this with 'α-white phosphorous', which refers to the substance characterized by its cubic crystal form. You cannot say it refers to the cubes, because each side of a cube has four atoms shared with the next cube.

A crude maxim is that many items used as examples of natural kinds are not sets and do not have members. Not substances, not diseases, and certainly not heat. Hence although kinds of living things may be regarded as sets, the candidates for being a natural kind are not, in general, sets. Indeed: there is *nothing* that they are in general.

A. A. Cournot 1851

The tradition of natural kinds exists only in the English language, and indeed before Quine it was wholly insular.[40] But the problem of natural versus artificial groups was European. A French contemporary of Mill's, A. A. Cournot (1801–1877), did turn his attention to the problem of natural groups. Logician, economist, probabalist and educator, he had the same instincts as Whewell and Mill, that philosophical logic and clear thinking might bypass the endless debates in natural history.

The tradition of natural kinds was not only insular but also empiricist. Cournot's book of epistemology was called *An Essay on*

[39] Mill, *Logic*, footnote to III. xxii. §6 added to the 6th edition of 1865, 585.

[40] There is no obvious way to translate 'natural kind' into French. *Genre naturel* and *espèce naturelle* were both used in the 1858 translation of Mill's *Logic*, once on the same page, but both make incomprehensible Mill's contrast between kinds on the one hand, and *species* and *genus* on the other. The translation of Quine used *espèce naturelle*. The translators of Putnam and Kripke followed suit. Cournot wrote about *genres naturels*, but he did not mean natural kinds. He meant natural as opposed to artificial genera; he believed that the species in use in the biology of his day were natural, and that the question of artificiality arose chiefly for higher ranks, starting with genera. Thus Cournot was addressing what I called the ontological, rather than the descriptive, problem about taxonomy.

the Foundations of our Knowledge and on the Characteristics of Philosophical Critique.[41] The first half of the title recalls the great books of the British Empiricists; the second half reminds us of Kant. That's Cournot for you. Neither rationalist nor empiricist, neither realist nor idealist, he had a thoroughly naturalistic view of knowledge as part of the human relationship with the world. In retrospect it seems curious that the English contributors said so little about causes in connection with natural kinds. Cournot's explanation of a natural group relies on the existence of underlying hidden causes that lead items to be grouped together—Putnam's hidden structures, if you will.

Author of a major philosophical work about probability, Cournot's master in matters of causation was neither Hume nor Kant but Laplace. Laplace did not address the metaphysics of causation, but the practical question of whether a phenomenon could be attributed to a cause or was simply a matter of chance. Cournot used the Greek constellations such as Cassiopeia as examples of artificial groups. They are convenient for navigators but the stars are grouped together by people, not nature. Cournot had translated William Herschel's *Treatise of Astronomy*. Herschel discovered the nebulae, what we now call galaxies. They furnished Cournot's contrasting example of a natural group. He used Laplace to argue that there must be some underlying cause that puts together the great clumps of stars identified as nebulae. 'Generic types and the classifications of naturalists give rise to remarks that are perfectly analogous [to the case of the nebulae]. A genus is natural, when the species of the genus have so many resemblances among each other, and by comparison differ so much from species that belong to neighbouring genera.'[42]

If different species fall under one (natural) genus, it must be highly improbable that a purely random assignment of individuals to the species, and of the species to the genus, should yield such a small distance between the species assigned to different genera. Referring to Laplace's theory of the probability of causes, he

[41] A. A. Cournot, *Essai sur les fondements de nos connaissances et sur les caractères de la critique philosophique* (Paris: Hachette, 1851).

[42] *Ibid.* 201. Darwin used exactly the same example to contrast with species explained in genealogical terms, *Origin* 397. Writing too soon after the *Origin* had been published, Cournot averred that we never would answer the question of the origin of species, but that we could tell on Laplacian grounds which groups of living things were natural groups. *Traité de l'enchaînement des idées fondamentales dans les sciences et dans l'histoire*, (Paris: Hachette, 1861).

Ian Hacking

concluded that the grouping 'cannot with any probability be attributed to the fortuitous play of causes that makes the types of organization of one species vary irregularly from that of another. There must be a bond of solidarity between the causes that constitute the species of a genus.'[43]

Cournot wrote as an astronomer and probability theorist rather than a naturalist. French taxonomists paid little attention to his characterisation of *genres naturels*, which was, in any event, to be upstaged by Darwin, admittedly rather more slowly in France than elsewhere. The English tradition of natural kinds ignored him altogether. The Laplacian probability of causes never entered the tradition, but it was Cournot, rather than the nineteenth-century empiricists, who saw that the idea of causality and of natural kinds had to be intricately intertwined.

1900–1970

The British doctrine of natural kinds was motivated by problems of natural history and the debates about natural groups that were still thriving in 1840. Those debates became obsolete. What with Venn and Peirce, the doctrine should never have entered the twentieth century. But other purposes were found for it. C. D. Broad tried to use natural kinds to analyse inductive inference. His exceptionally fine essay was the first to make clear that: 'The notions of permanent substances, genuine natural kinds, and universal causation are parts of a highly complex and closely interwoven whole and any one of them breaks down hopelessly without the rest.'[44] Broad was not able to fix any of the three to his satisfaction, and he confessed that he published the paper because he could not bear to think about this network of notions any more. Some readers will judge that this is a brilliant *reductio ad absurdum* of the idea, that you can elucidate this triangle of notions in an empiricist way.

The two great empiricist philosopher-logicians of the twentieth century, Quine and Russell, agreed completely about the idea of a

[43] Cournot, *Essai*, p. 202, 204. His words were *lien de solidarité*, a concept that he does not explain very clearly. But it is not idle to associate it with Putnam's concept of a hidden structure underlying a natural kind.

[44] C. D. Broad, 'On the Relation between Induction and Probability', Reprinted from *Mind* 27 (1918), 389–404; 29 (1920), 11–45, in *Induction, Probability, and Causation: Selected Papers by C. D. Broad* (Dordrecht: Reidel, 1968), 1–52, on 44.

natural kind. They thought of it as a sort of epistemological crutch for getting started in the world. 'The existence of natural kinds', wrote Russell, 'underlies most pre-scientific generalizations, such as "dogs bark" or "wood floats".' He gave a standard Millian explanation of the idea of natural kind and its uses. But in the end he had to 'conclude that the doctrine of natural kinds, though useful in establishing such pre-scientific inductions as "dogs bark" and "cats mew", is only an approximate and transitional assumption on the road to more fundamental laws of a different.'[45]

Quine, with his gift for giving the gist in the smallest number of words, gave us my favourite five-word characterization of natural kinds: 'functionally relevant groupings in nature'. Nevertheless he concluded his famous essay by repeating Russell's thought. 'In general we can take it as a very special mark of the maturity of a branch of science that it no longer needs an irreducible notion of similarity and kind.' Indeed the disappearance of this notion is, 'a paradigm of the evolution of unreason into science'.[46]

High Noon: Kripke and Putnam

My topic is dawn and twilight, not the heady days of the 1970s when Saul Kripke and Hilary Putnam did so much to give sense and use to the idea of a natural kind. Analytic philosophy is directed more at semantics than at nature. Hence causal theories of reference for natural-kind *terms* are often called simply the Kripke-Putnam theory. This is correct; indeed Putnam acknowledged a debt. 'Kripke's work has come to me second hand; even so, I owe him a large debt for suggesting the idea of causal chains as the mechanism of reference'.[47] It is seldom noticed that Kripke's and Putnam's theories of natural *kinds* are very different. Kripke single-handedly brought talk of essence back to life. This was an amazing feat. One would have thought that it was dead as a doornail in English-language analytic philosophy once Locke had savaged it with such relentless irony in that 'immortal Third Book'. To

---

[45] B. Russell, *Human Knowledge: Its Scope and Limits* (London: George Allen and Unwin, 1948), 335, 461–2.

[46] Quine, 'Natural Kinds', *op. cit.* note 6, 126, 138.

[47] H. Putnam, 'Explanation and Reference', in *Mind, Language and Reality: Philosophical Papers*, Vol. 2 (Cambridge: Cambridge University Press 1975), 198.

paraphrase Mill quoted above, Kripke brought it back from 'some remote fastness' and restored it to pride of place in 'open country'.

Putnam had almost no part in the restoration of essence. At the beginning he was quite willing to present his ideas as parallel to Kripke's, and to use the myth of baptism as giving a reference to natural kind words. Yes, he did *mention* essences—in scare quotes. When he put forward his fundamental idea that natural kinds have 'the same general *hidden structure*', he added, between parentheses, '(the same "essence", so to speak)'.[48]

Kripke and Putnam both formulated semantic theories of natural-kind terms. As a rule of thumb we may say that Kripke's theory of natural kinds was logico-metaphysical, with essence at its core, while Putnam's was empirico-logical, with hidden structure at its core. It was some time before Putnam came to realize that Kripke meant every word he said, and hence his statement became more and more clearly differentiated from Kripke's approach. I point this out in some detail elsewhere.[49]

TWILIGHT

Proliferation

The work of both men produced a rich sub-discipline of philosophy. There have been endless debates and numerous criticisms.[50] Yet despite the initial enthusiasm, by 2006 we are left with a great many almost unrelated research ideas about natural kinds. I shall not detail here how we got from high noon to here. I shall briefly summarize a situation that will be well-known to many readers. The present situation is scholastic, in several senses. The

[48] H. Putnam, 'The Meaning of "Meaning" ', in *ibid.*, 215–271, on 235.

[49] I. Hacking, 'Why Putnam's Theory of Natural Kinds is not the same as Kripke's', to appear in *Principia: Revista Internacional de Epistemologica* (Florianopolis, Brazil). 'Hidden Structure and Natural Kinds', to appear in the Library of Living Philosophers ('Schillp') volume dedicated to Putnam.

[50] One of the most vigorous recent critiques is J. Laporte, *Natural Kinds and Conceptual Change* (Cambridge University Press, 2004). It contains thorough references to thirty years of debate. A decade earlier T. E. Wilkerson, offering a modest essentialism, provided ample references in *Natural Kinds* (Aldershot: Avebury, 1995), with an update, 'Recent work: Natural kinds' *Philosophical Books* **39** (1998): 225–233.

great schoolmen were deeply caught up in questions about general terms and classification. In this sense, the connotation of 'scholastic' is positive. Second, they argued exquisitely about the finest points. The present debates about natural kinds are reminiscent of those noble hours of the late middle ages. But there is a third connotation that I fully intend. 'Scholastic' suggests an inbred set of degenerating problems that have increasingly little to do with issues that arise in a larger context.

This is not to say that real problems do not abound. I say only that discussing them in terms of natural kinds does no good at all. It is an optional add-on. Not an empty add-on, because the term 'natural kind' now carries a lot of baggage with it, and a lot of mutually incommensurable theories that I am about to list. So speaking of natural kinds turns real difficulties into unnecessary confusions. Whewell made 'kind' a free-standing logical term intending to solve or evade problems about natural groups. The species remain a much-debated question in systematics, as do the higher taxa. But to discuss them in terms of natural kinds today is to spill ink. Take any discussion that helps advance our understanding of nature or any science. Delete every mention of natural kinds. I conjecture that as a result the work will be simplified, clarified, and be a greater contribution to understanding or knowledge. Try it.

Definitional confusion

Usually encyclopaedia articles summarize fairly standard recent knowledge. One may disagree with an entry because one is a rival expert or because one mistrusts the ideology implicit in the article, but in general we accept encyclopaedias as authoritative. I do not wish to be invidious, but it is convenient to use the article 'Natural Kinds' in the *Routledge Encyclopedia of Philosophy* to illustrate the confused state of the philosophy of natural kinds. An obvious response is that the article is defective, and that happens. Agreed. I offer no more than an illustration before turning to the details of proliferation.

> Objects belonging to a natural kind form a group of objects which have some theoretically important property, or properties, in common. Standard examples of natural kinds include biological species such as rabbits, oaks and whales, chemical

elements and compounds such as oxygen, carbon and aluminium, and stuffs such as salt, wool and heat.

I personally favour a mundane, rather than cosmic (Quine's word again) notion of natural kinds, and in that mood am content with these examples. But what is said about them is troubling. None of the kinds called biological species are in fact biological species.[51] They are respectively a genus, an order and a family. *Oak* is the genus *Quercus* (and some oaks are of the genus *Linocarpus*). *Whale* in biology denotes most of the order *Cetacea*. And *Rabbit* may be any of the long-eared burrowing animals of the family *Leporidae*. As noted below, there was a brief moment in molecular biology when one expected to find exactly what members of a given biological species have in common. Considered opinion now has it that there is nothing that all and only members of a given species have in common.

As for substances rather than kinds of living things, what determines that salt is filed as a stuff rather than a chemical compound? Heat is a *stuff*? What wool does the author have in mind? The dense, soft, often curly hair forming the coat of sheep and certain other mammals, such as the goat and alpaca, consisting of cylindrical fibres of keratin covered by minute overlapping scales and much valued as a textile fabric? Or what most of us think of, a textile fibre made from raw wool, or perhaps a material or garment made of this textile?

The New Essentialism

Keith Donnellan wryly observed that a weird thing about the discussions of Kripke and Putnam was that they almost always took examples from common English, and not the innumerable technical names actually introduced into the sciences, in order to refer to newly discovered or understood kinds of things.[52] Water, lemons, heat, gold. Brian Ellis's Scientific Essentialism fully rectifies that. It emphasizes three types of natural kinds. *Substantival natural kinds* include elements, fundamental particles, inert gases, sodium

---

[51] The same point was made long ago in J. Dupré, 'Wilkerson on Natural Kinds', *Philosophy* **64** (1989), 248–251. Wilkerson modified his account in the light of the criticism.

[52] K. S. Donnellan. 'Kripke and Putnam on Natural Kind Terms', in C. Ginet and S. Shoemaker (eds.), *Knowledge and Mind: Philosophical Essays*, (New York: Oxford University Press, 1983), 84–104.

salts, sodium chloride molecules, and electrons. *Dynamic natural kinds* include causal interactions, energy transfer processes, ionizations, diffractions, $H_2 + Cl_2 \Rightarrow 2HCl$, and photon emission at $\lambda = 5461\text{Å}$ from an atom of mercury. *Natural property kinds* include dispositional properties, categorical properties, and spatial and temporal relations; mass, charge; unit mass, charge of $2e$, unit field strength, and spherical shape.[53] Species are *not* natural kinds, in this philosophy.

I fully respect, although I do not share, the anti-Humeian metaphysic that motivates Ellis's essentialism. But why should we say that *these* are the natural kinds? These various kinds of entities are all named in my undergraduate textbooks published in the 1950s. Very well, assert that these kinds have essences, if that helps you understand their agency. But should one not be worried that no science textbook of 1956 or 2006 ever mentions the essence of photon emission or anything else?

These kinds of item have stood up well, these fifty years, but why, I repeat, call them the natural kinds? That rhetorical add-on may give lustre to a new theory by hooking it up with an old tradition, but it adds not one jot of content. As stated in the first thesis at the start of this paper, stipulative definition, humpty-dumpty style, is always possible, and Ellis can define 'natural kind' as he will, but what good does it do?

Note that the causal theory of reference derived from Kripke and Putnam plays no role at all in the New Essentialism. Of course 'argon' continues to denote argon, but that goes without the theory. Kripke's motivation for introducing essences has entirely disappeared from the New Essentialism; we are left with bare essences. Those who want natural kinds to underwrite inductive inference must also be sorely disappointed, for we make inductive generalizations (*pace* Popper) about poplars, possums and potatoes, which are said not to be natural kinds, as readily as we do for potassium.

Michael Ghiselin's evolutionary biology

For some forty years Michael Ghiselin has been urging that the species (Mill's *horse*, say, *Equus caballus*, genus *Equus*) are

---

[53] B. Ellis, *Scientific Essentialism* (Cambridge: Cambridge University Press, 2001), 56.

individuals.[54] The individual horse, Black Beauty, is a *part*, and not a member, of this individual. On the other hand the ranks—species, genus, and the rest—are natural kinds.

The first assertion is Ghiselin's way of saying that the unit of natural selection and hence of evolution is the species, an individual that evolves. Moreover, there are no exceptionless laws of nature about horses. His second assertion means that there are biological laws of the form 'every species is so and so'. These are truly important propositions.

Ghiselin's logic of the species is mereological, while his logic of the taxa is set-theoretic. Some of us, starting perhaps with an old paper by Philip Kitcher,[55] suspect that such theses can be stated using a more conventional logical approach. Perhaps not. Either way, the biological substance of Ghiselin's proposals will be left intact. What they do not need is the rhetorical add-on about natural kinds.

Ghiselin fumes at John Dupré's doctrine of promiscuous natural kinds, which urges that several incommensurable modes of classification may be used for classifying living things. Ghiselin is not dogmatic; he does not say that the taxonomy he favours is right in all respects. But if it is not right, then another one *is* right. And whatever it is, it will be *the* fulfilment of Darwin's insight. It does not help, in my opinion, to conduct this polemic using the rhetorical label 'natural kind'.

Developmental Cognitive Science

An unexpected marriage of Chomskyian cognitive science and Piagetian psychology holds that many of the abilities, which children acquire early in their lives, are enabled by innate mental modules. One of these is a natural-kind module. It enables children at an early age to begin to classify, to generalize over classes, and to pick up common names for the classes. There is the additional thesis that children act as if the classes, for which they are innately primed, have essences. It is not asserted that metaphysical essences exist, but that children, and later on, adults, act as if they did.

[54] M. Ghiselin, 'On Psychologism in the Logic of Taxonomic Controversies', *Systematic Zoology* **15** (1966), 207–215. *Metaphysics and the Origin of Species* (Syracuse, N.Y.: State University of New York, 1997).

[55] P. Kitcher, 'Species', *Philosophy of Science* **51** (1984): 308–333.

Natural Kinds: Rosy Dawn, Scholastic Twilight

Frank Keil is a leading worker in this field.[56] In a brilliant synthesis of cognitive science, Aristotelian scholarship, history of systematic biology and cross-cultural anthropology, Scott Atran has done most to advance the idea of a living-thing module for natural kinds.[57]

Atran thinks that Aristotle's accounts of living things pretty well reflects what he calls folk-biological concepts, which are universal in the human race. But he also studies, in rich historical detail, what he calls 'the scientific breakaway', in which natural history and then systematic biology replaced all those concepts by a nested hierarchy of taxa. It is hardly an exaggeration to say that none of the 'natural kinds' in folk biology coincides, except in rough outline, with any species or genus of systematics. Atran rightly concludes that the idea of natural kinds, pertaining to every kind of thing, is on the way out:

> The conception of 'natural kind', which supposedly spans all sorts of lawful natural phenomena, may turn out not to be a psychologically real predicate of ordinary thinking (i.e. a 'natural kind' of cognitive science). It may simply be an epistemic notion peculiar to a growth stage in Western science and philosophy of science.[58]

My only disagreement is that the conception of a natural kind never had a role in Western science—it was peculiar only to a growth stage in English-language philosophy of science.

Biological species

When Kripke first published, current scientific folklore held that in the next few years molecular biology would discover in the DNA of a species the necessary and sufficient conditions for being of that species. Hence one blithely talked of the essence of tigers and lemons. We would move from the phenotype to the genotype and on to the essence. Within twenty years the folklore had been

[56] F. C. Keil, *Semantic and Conceptual Development: An Ontological Development* (Cambridge, Mass.: Harvard University Press 1979). *Concepts, Kinds and Cognitive Development* (Cambridge, Mass.: MIT, 1989).

[57] S. Atran, *Cognitive Foundations of Natural History: Towards an Anthropology of Science* (Cambridge: Cambridge University Press, 1990).

[58] S. Atran, 'Folk Biology and the Anthropology of Science: Cognitive Universals and Cultural Particulars,' *Behavioral and Brain Sciences* **21** (1998), 547–569, with discussion and replies until 609, on 569, note 16.

superseded. There are no necessary and sufficient conditions for being a tiger or a lemon. No essence. Terence Wilkerson presented a far more modest essentialism than Brian Ellis, but holds that the biological species as we know them are not natural kinds.[59] It remains to be said that projects are afoot to identify genetic 'barcodes' for recognizing species, but these will be contingent markers, not essences that explain why the Toco Toucan has an enormous bill. They will be more like DNA fingerprints, which may identify Elizabeth II but do not define her.

Species die hard as the basic paradigm for natural kinds. Rachel Cooper ran through many of the competing accounts of natural kinds, not mentioning the New Essentialism.[60] Without herself taking any position on the definition of natural kinds, she regarded Wilkerson as an outlier, and concluded that most theories about natural kinds still count biological species as natural kinds. John Dupré, for example: 'there is no reason why the account of species currently offered [in classificatory systematics] should preclude their being modestly natural kinds'.[61] Memories die hard. Biological species have long served as paradigms of natural kinds—they are (one feels) natural kinds if anything is. Witness the definition in the *Routledge Encyclopedia of Philosophy*, with its oaks, rabbits and whales all called species.

Richard Boyd's Homeostatic Property Cluster Kinds

One of the most innovative approaches to species after Putnam is due to Richard Boyd.[62] The idea of homeostasis was introduced in the 1920s to describe human metabolism. It was taken over by

[59] Wilkerson, *op. cit.* note 50.

[60] R. Cooper, 'Why Hacking is Wrong about Human Kinds', *British Journal for the Philosophy of Science* **55** (2004), 73–85. Her point was to show that I was wrong about what I used to call human kinds. I do not protest her argument, but go further back. There is, if possible, even less of a class of human kinds than there is of natural kinds. See Ian Hacking, 'Kinds of People: Moving Targets', forthcoming in *Proceedings of the British Academy*.

[61] Dupré, *Humans and Other Animals*, *op. cit.* note 2, 97.

[62] R. Boyd, 'What realism implies and what it does not', *Dialectica* **43** (1989): 5–29; 'Anti-foundationalism and the enthusiasm for natural kinds', *Philosophical Studies* **61** (1991): 127–148; 'Homeostasis, species and higher taxa', in R. Wilson, *Species: New Interdisciplinary Essays* (Cambridge, Mass.: MIT, 1999), 141–186.

cybernetics together with the idea of positive and negative feedback. A system is homeostatic if it has an equilibrium state, and whenever it strays too far from that state, there are causes in the way of feedback that tend to restore equilibrium. Boyd adapts this concept for species and higher taxa, combining it in an unusual way with the Darwinian model of selective pressure..

In his analysis, kinds, and in particular species, are groups that persist in a fairly long haul. The properties that characterize a species form a cluster. No distinctive property may be common to all members of the species, but the cluster is good for survival. The species is in equilibrium in the sense that descendants that diverge too far from the cluster of properties die out or form a new group. Species thus endure thanks to a network of causes that produce stability of a homeostatic sort. That is, when members of successive generations of a species deviate too far from an earlier prototype, they either survive, and the phenotype of the species gradually changes, or else they die, leaving the surviving majority to keep the species prototype intact.

Boyd would like to extend his idea in many directions, including epistemological concepts. Knowledge, in his opinion, is a homeostatic property kind. It is thus a natural kind of epistemological state, in a way in which (if I understand him) belief is not. Note how the notion of homeostatic property cluster kinds may extend to kinds never before included in the pantheon of natural kinds, while it is not useful for old-time natural kinds such as phosphorus. Like many other kinds of stuff filed as natural kinds, phosphorus is highly unstable, the very opposite of homeostatic, which is why it is almost never found free in nature. It is made in factories, and preserved, in one or more of its allotropic forms, by artifice.

Induction

There are calls for natural kinds motivated by a wish to understand the philosophical problem of induction. I shall make four observations.

(a) As C. D. Broad, quoted above, may have been the first to say in detail, substance, kinds, and causation 'are parts of a highly complex and closely interwoven whole'. It is remarkable how few are the authors who have seriously examined these as parts of a whole. David Wiggins has doggedly treated the first two under the rubric of substance and sameness. Ruth Millikan has proposed 'a common structure for individuals, stuffs and real kinds'.[63] There is a depth in the problems addressed by Wiggins and Millikan that seems lacking in most philosophising about induction without discussing substance.

(b) 'In induction', as Quine observed, 'nothing succeeds like success'. Hence the kinds—I follow Quine and drop the 'natural'—are the predicates that we come to use, and will regularly revise. Calling them natural kinds adds nothing. To call them kinds and speak of sameness of kinds is to speak with William James' pragmatism.

(c) Howard Sankey urges that: 'Inductive inference in science is rationally justified because of the existence of real, natural kinds of things, which are characterized as such by the essential properties which all members of a kind must possess in common.'[64] Are inductive generalizations about rabbits, oaks and whales not rationally justified, since biology now teaches that they lack such essential properties? As a matter of fact, induction works best with artefacts, because that is exactly what artefacts are, things that fairly reliably do what we want them too. That makes rather a hash of the popular distinction between 'natural' and 'artefactual' kinds, if one imagines that a chief task of natural kinds is to underwrite induction.

(d) To use a name for any kind is, among other things, to be willing to make generalizations and form expectations about items of that kind. That is a primary lesson to be drawn from Goodman's new riddle of induction.[65] The very words 'kind' and 'generalize' have the same roots—a story well narrated by the *OED*.

[63] D. Wiggins, *Sameness and Substance* (Cambridge, Mass.: Harvard University Press), 1980. *Sameness and Substance Renewed* (Cambridge: Cambridge University Press, 2001). R. B. Milikan, 'A Common Structure for Concepts of Individuals, Stuffs, and Real Kinds: More Mama, More Milk, and More Mouse', *Behavioral and Brain Sciences* **21** (1998), 55–65, with discussion and replies until p. 100. *On Clear and Confused Ideas: An Essay about Substance Concepts* (Cambridge: Cambridge University Press, 2000).

[64] H. Sankey, 'Induction and Natural Kinds', *Principia: Revista Internacional de Epistemologia* **1** (1997), 239–254.

[65] Indeed it is essentially the first sentence of my paper 'Entrenchment', in D. Stalker, (ed.), *GRUE! The New Riddle of Induction* (Chicago: Open Court), 193–223.

Natural Kinds: Rosy Dawn, Scholastic Twilight

Laws of nature

Natural kinds are invoked to explicate the idea of a law of nature, or vice versa. One message of *Fact, Fiction, and Forecast* was that the connection was quite strong, so that they stood or fell together. Goodman judged that they fell. For quite other reasons, laws of nature are under a lot of pressure right now. In physics there is Nancy Cartwright's doctrine that they are all false, and only approximations are true.[66] There is Bas van Fraassen's thesis that 'no philosophical account of laws of nature does or can succeed.'[67] Symmetry is where the action is, he argues, rather than law. This opinion is at present asserted, if not so fervently, by many other philosophers of science. Turning to evolutionary biology, Ghiselin held that laws of biology apply to taxa, such as *species,* rather than to items that fall under the concept 'species'. This has certainly been contested.[68] The concept of a natural kind contributes nothing to those debates about fundamental issues, and they had better be settled before one imagines that the concept of a law of nature can be routinely invoked to explain the idea of a natural kind.

Philosophical purists no longer favour the concept of generalizations that are true for the most part. Yet these are good for predicting, and are very often the very propositions that earlier generations counted as laws of nature. I myself am no purist. Whewell was, in my opinion, on the right track when he said that a kind is a class denoted by a common name about which there is the possibility of general, intelligible and consistent, and probably true assertions. Thus there are plenty of law-like generalizations about both substances and species. Exposure to sunlight or heat changes white phosphorus to red, which neither phosphoresces nor ignites spontaneously in air. The Hyacinth Macaw (*Anodorhynchus hyacinthinus*) eats only the white fruit inside two kinds of very small palm nuts (*Suagrus commosa* and *Attalea funifera*); first animals such as peccaries, capybara or cattle must eat the green husk in order that the strong bill of the macaw can crack the nut underneath.

---

[66] N. Cartwright, *How the Laws of Physics Lie* (Oxford: Clarendon, 1983).

[67] B. van Fraassen, *Laws and Symmetry* (Oxford: Clarendon, 1989), vii.

[68] For example, by Dupré, *Humans and Other Animals, op. cit.* note 2, 108; not even the Hardy-Weinberg law will do the trick.

Ian Hacking

Many philosophers prefer to insert Latin tags here, *mutatis mutandis* or *prima facie*, or they speak of *ceteris paribus laws*. We do not bother to do so in plain English. The tags reveal that the philosophers want a definite class of such laws. Latin is the polite printed form of hand-waving. In fact there is no more a specific class of such laws than there is a specific class of natural kinds.

My two examples deliberately refer to facts and kinds of things about which few people are well informed. They present what for many readers will be new classifications and new knowledge. I like to use the label 'mundane' for the several kinds of minerals, animals and vegetables mentioned. Mundane kinds as opposed to cosmic ones.

In a first glance at natural kinds many years ago, I used the dictionary for a random sample of natural kinds, and by chance found a surprising number of mundane kinds whose names began with 'stone-'.[69] Few were minerals; instead I listed various kinds of plant, fish, fruit, bird, insect, algae and the stone-lily, which is an invertebrate marine animal. My favourite mundane kind is mud. Mud is a functionally relevant grouping in nature. It is familiar to parents scrubbing children's clothing, to football players, and to ditch-diggers in damp climes. Not scientific? I used to work around oil rigs in the Prairies. We always had a mud-engineer on the site, a man not ashamed to have that title, *mud engineer*, on his business card. (For current rates of offshore pay, from Aberdeen to Yemen, consult the Internet.)

What arrogance will insist that the ploughman is not in touch with nature's kinds, mud and dung?

Conclusion

Although one may judge that some classifications are more natural than others, there is neither a precise nor a vague class of classifications that may usefully be called the class of natural kinds. A stipulative definition, that picks out some precise or fuzzy class and defines it as the class of natural kinds, serves no purpose, given

[69] Ian Hacking, 'A Tradition of Natural Kinds', *Philosophical Studies* **61** (1991), 109–126. My long-postponed book, *The Tradition of Natural Kinds* (forthcoming with Cambridge University Press), tells more about the dawn, treats the high noon of Kripke and Putnam in respectful detail, and moves on to the present twilight.

that there are so many competing visions of what the natural kinds are. In short, despite the honourable tradition of kinds and natural kinds that reaches back to 1840, *there is no such thing as a natural kind.*

The Moral Use of Technology

JAMES GARVEY

There is a well-worn example—well worn in some circles, anyway—of what you might think of as racist bridges.[1] Robert Moses, the celebrated New York architect, designed unusually low bridges over the roads from Long Island to Jones Beach. The bridges were designed in this way so as to prevent the poor and predominantly black locals from travelling to the beach by bus—the affluent, white car owners can slip under them with ease. The bridges prevent certain members of the community from enjoying the beach as surely as a phalanx of clansmen. Perhaps the strongest moral drawn from the story is the claim that the bridges are political. The objects themselves are imbued with dubious values.

I am not sure how much of the story to credit or exactly how to interpret the conclusion, but I do at least find it suggestive.[2] If moral questions might be raised by something as innocuous as a bridge, it is possible to start worrying about the effects of other everyday artefacts, the technological stuff we are in contact with daily. You might worry about the morally acceptable uses of all of the stuff; you might wonder whether or not you are making a moral mistake in using technology in some way rather than another or in using it at all. Let us try to pin some of this down.

Is technology neutral, a neutral means to whatever ends we have in mind, or is it, instead, somehow imbued with moral and political value, a kind of autonomous force which brings about its own ends? This question will make a little more sense once we consider the two answers it suggests. The answers need names, and I'll go along with at least some of the literature and call the former 'the instrumental view' and the latter 'the autonomy view'. In this paper, what I really want to determine is whether or not the answers can be brought to bear on a second question about technology, a slightly more human and certainly smaller question, a question which might even matter: how should we think about the

---

[1] Langdon Winner, 'Do Artefacts Have Politics?', *Daedalus* **109**, No. 1, 1980, 121—36.
[2] For a generally clear-headed, sceptical discussion of the example, see Bernward Joerges, 'Do Politics Have Artefacts?', *Social Studies of Science,* **29**, No. 3, 1999, 411—431.

241

James Garvey

moral dimension of mundane technology, in particular, what is the right way to use mundane technology? In a time-honoured philosophical tradition, I will tell you about both views, try to show that neither one is much help, and then point to the beginnings of an alternative.

You might think, at the outset, that you are owed a definition of technology, in particular, a definition of mundane technology. You are not going to get one from me, at least not a formal one. Definitions tend to settle debates without breaking sweat, and it might be best to avoid settling on one for as long as we can. I'll give an example of this sort of thing which will also convey, in a backhanded way, the kinds of definitions of technology operative in at least some thinking in this neighbourhood.

Kline identifies four general uses of the term 'technology'.[3] The most common use, he says, equates technology to hardware, manufactured artefacts, shoes and ships and ceiling wax, but not cabbages or kings. The second use of 'technology' is manufacturing hardware, the equipment used to manufacture objects—this might include people, the human cogs in the wheels of production. The third sense identified equates technology with techniques, know-how or methods used in the production of objects. The fourth conception of technology, the one favoured by Kline, takes it that technology is understood as sociotechnical systems of manufacture and use: a system combining both the production of hardware and the people who use it, as well as other elements required to put the hardware to use.

The point of at least the first three definitions can seem more or less obvious. The word does get used to pick out stuff or the manufacture of stuff or the know-how required to manufacture stuff, but you can wonder about the fourth definition. The point of the fourth sense of the word 'technology', for Kline, seems to be that an adequate definition has to include more than just the immediately obvious material stuff, because the tasks performed with the use of technology require more than the material stuff. What's required to drive a car is not just the car, but the driver and her knowing how to drive, car manufacturing plants, roads, gas stations, laws, rules of the road and, in general, the many social and technological structures underpinning the manufacture and use of cars. Without all this, a car could not be driven: so all of this is car technology. You can be rightly suspicious of Kline's thinking here,

[3]   Stephen J. Kline, 'What is Technology?', *Bulletin of Science, Technology & Society*, 1985, 215–18.

but for now just notice the implications of the definitions for our general question about technology.

Without getting too far ahead of ourselves, you can probably already see that plumping for something along the lines of his first, second or third definitions—technology as tools, the manufacture of tools, and the knowledge required to manufacture tools—will land you with the view that technology is a neutral instrument or set of instruments. Opting for something closer to definition four, technology as a sociotechnical system, will make technology seem like something more than a neutral means, something closer to a burgeoning, autonomous force. You can also probably already see the sense in which thinking about technology as a neutral means can actually lead to the view that technology has to be something more than just a neutral means. Kline has done some thinking about the tasks that we perform by using technology (instrumentalism) and come to the conclusion that we cannot understand those tasks without understanding the sociotechnical systems which underpin them (something near autonomy). I'll try to make both points more clear in the following sections.

If you allow that I am understandably reticent about settling on a definition of technology, you might still insist on some explanation of the word 'mundane'. There is a distinction, an admittedly barely rough and ready one, between mundane and exciting technology. Moral and political philosophers of a practical bent are everywhere exercised by questions concerning exciting technology: nuclear weapons, unravelling the human genome, GM crops, cloning, stem cells, senescence technology, nanotechnology, thinking machines, virtual reality, and on and on. The stakes are thought to be high, and probably they are. If we make a moral mistake about nuclear weapons, it's the end of the world.

Exciting questions about exciting technology are, of course, pressing and worth our attention, but I want to avoid thinking about them. I want to avoid such questions mostly because they scare me but also because certain aspects of exciting technology can be a little distracting, can nudge us away from the point of our original question about the moral dimension of technology generally. It also nearly goes without saying that a consideration of exciting technology would drag us away from our second question, which concerns just mundane technology and the moral implications of its use. Other and better minds are covering the exciting territory well enough. Anyway, if we make a mistake in our thinking about mundane technology, it's not the end of the world.

James Garvey

If formal definitions are not on the cards, we can make do with ostensive ones. Mundane technology includes electronic gadgetry like mobile phones, computers, cameras, camcorders, DVD players, as well as heftier items like cars, boats, buildings and bridges, and other similar things. These things are all tools, but if you want to think of them as something more, if you want to join Kline and think of them as objects embedded in sociotechnical systems or even as sociotechnical systems, you are welcome to do so. It is hard to find the objects themselves very interesting, primarily because they are pervasive, more or less everywhere. Their being everywhere, though, should make them a little interesting. Mundane technology has more effects on us than might be imagined—just because we are surrounded by it and interact with it from moment to moment—and some of those effects might be moral or political in flavour. The little drips and drabs can add up. You can wonder about all those little effects, and you can look to the instrumental view or the autonomy view of technology for help in thinking about them, which is what we'll now do.

1. The Instrumental View

The thesis of instrumentalism is expressed in contemporary thinking mostly by opponents. Those who consider it tend to mention it briefly, perhaps pointing out that it is 'the common sense' or 'default' position before quickly moving on to something else. Feenberg, who certainly has moved on to something else, says that instrumentalism 'offers the most widely accepted view of technology. It is based on the common sense idea that technologies are "tools" standing ready to serve the purposes of their users. Technology is deemed "neutral", without valuative content of its own'.[4] Arguments for the view are a little thin on the ground. The instrumental view is supposed to be the older view. Many point to Bacon in an attempt to identify its beginnings.

What you find in Bacon, however, is nothing like an argument for the view that technology is neutral, but something closer to a vision or a hope. Nature, for Bacon, consists in a chain of hidden causes. Human beings cannot break the chain, but coming to understand the chain is, in itself, a kind of power over the natural world. Nature, Bacon sloganizes, is only overcome through obedience.

[4] Andrew Feenberg, *Critical Theory of Technology* (New York: Oxford University Press, 1991), 5.

There is, in all of this, certainly more than just a whiff of the view that technological innovation is merely a means to our ends, perhaps because the emphasis is clearly on the ends. The hope, in harnessing nature, is 'an improvement of the estate of man ...an enlargement of his power ...so as to provide man with his bread, that is, with the necessities of human life.'[5] Bacon is after more than just bread. His claim is that technology is our way of making things better for ourselves, getting what we need not just to live— scrabbling around like animals would accomplish that—but to improve the human condition. Bacon thinks that technological and scientific innovation delivers not just bread but a host of technological delights. A consideration of what he calls 'the true state of Salomon's house' will tell you as much.[6]

The technological utopia he envisages includes caves where artificial metals are produced, alongside wondrous 'coagulations' which prolong human life and cure disease. There are great houses where whole weather systems are replicated—complete with snow, hail, rain, lightning, and, rather oddly, where the generation of frogs and flies is undertaken. Mechanical devices churn out paper, silk, 'dainty works of feathers of wonderful lustre'. Furnaces hum along, imitating the sun's heat. Engines produce astonishing motions. Terrifying weapons are stockpiled. 'These are (my son) the riches of Salomon's House.' Not bad, you might think, but there follows a slightly disturbing discussion of various emissaries of the state, who disguise themselves, sail off to other lands, and steal the books, artefacts and experimental methods of other nations. Salomon's house has excellent riches in it, but still more excellent riches are required, and a little burglary is a small price to pay for them.

Is it going too far to suggest that something along the lines of Bacon's near delirious fascination with the fruits of technology and scientific enquiry is a foundation for the instrumentalist view? Bacon's emphasis, it seems clear, is on what the study of nature can do for us, and he certainly has high hopes. His eyes are fixed almost entirely on the intended ends. Technology is nothing more than instruments, devices, maybe knowledge used to achieve something nearly miraculous, but it's the something nearly miraculous that gets attention. It might be right to say that this is the default view.

[5]   Francis Bacon, (Peter Urbach and John Gibson, Ed., trans.) *Novum Organum* (Chicago and La Salle: Open Court Publishing, 1994), 293.
[6]   Francis Bacon, (Jerry Weinberge, Ed.) *New Atlantis* and *The Great Instauration*. (Wheeling, IL: Harland Davidson, Inc, 1989), 71–83.

245

Bacon shows us that if our eyes fall comprehensively on the ends we hope to achieve, our default conception of technology is merely a means for getting something else.

It's not just Bacon. You can find this emphasis on the ends in lots of places, and in more or less every case the technological means just fall into a neutral background. Miller, writing in 1803, tells us that the fact that mechanical arts 'have contributed, and will probably yet contribute, in a considerable degree, to the abridgement of labour, to the convenience and profit of artists, and to the excellence and beauty of manufacture, is too obvious to require particular explanation'.[7] His essay mentions chemistry, spinning cotton, steam engines, bridges, mills, carriages, printing, tanning, dyeing, metalwork and much else, and in each case the objects or processes themselves get only a moment's notice. What matters are the marvellous results. You can read a little American post-war rhetoric, talk of the march of progress from the 40s through to the 70s, and come to the conclusion that not much has changed. You might still be able to hear it today too.

Noticing expressions of instrumentalism, if that is what they are, is not to say that there have been no dissenting voices since Bacon. We have had plenty of Shelleys, and Huxleys, and Orwells; we've had Romantic disillusionment, not least with Rousseau; and we've had Luddites. You might wonder whether those dissenting voices are themselves thinking just about the ends. The point is, though, that if we are after arguments for the view that technology is a neutral instrument, what we usually find is something closer to a general failure to see anything but the ends we intend. What we get are visions and hopes, with technology itself falling neutrally to one side.

Can we do any better than this? Are there arguments for the instrumentalist view of technology? We'll have a look at two of them. The first takes it that means, including technological means, have moral value only derivatively, given the ends at which they are directed. The second infers neutrality from the functional promiscuity of technology. The two arguments might really only be variations on a single theme, but I'll treat them separately anyway, as I can't find a way to articulate a single argument which covers the ground as thoroughly as dividing things up.

Reflection on the nature of means and ends can land you with the view that means acquire value only with reference to an end.

---

[7] Samuel Miller, 'The Mechanic Arts', *A Brief Retrospect of the Eighteenth Century* (New York: T. & J. Swords, 1803).

Without an end in view, the means themselves are neutral, without value. This is roughly where Aristotle ended up—or, rather, began—and probably philosophy has largely followed him. In the first lines of *Nicomachean Ethics*, Aristotle claims that every art, investigation, action or pursuit it thought to aim at some good.[8] There is a sense in which our aims would be circular and ultimately pointless if they were just directed at each other, at other minor ends. If this is done for that, and that for something else, and that something else for another something, which might turn out just to be the original goal, with no supreme or general goal in mind, then we are going around in breathless, little circles. All of the little goals must themselves aim at some supreme goal, which Aristotle calls 'the good' or 'happiness', otherwise human beings are left with a potential infinity of minor aims and no chance of an ultimate point to action.

You need not buy into talk of 'the good' to take Aristotle's point here. The important claim is that our actions have no value in themselves but only acquire value if they aim at something which itself has value. Aristotle, of course, is talking about minor ends, like wealth, and he is arguing that such things only have value if pursued for something good in itself, say, happiness. Reasonable people have thought since that there might be more than one thing with intrinsic value, or at least more than one thing worth pursuing for its own sake, or, at the very least, lots of things which are simply worth pursuing. You can swap this pluralist view for Aristotle's talk of 'the good' if you like. Either way, you are left with the claim that undertaking an action has value only if there is a valuable end in view. The value of the means, it seems a reasonable way of putting it, is parasitic on or derivative of the value of the goal or end of the action.

Grant offers a nice statement of this line of thinking translated into talk of the value of technology. He starts by noting that value is of two kinds: intrinsic or instrumental. Either a thing is valuable in itself or it enables us to get something which is valuable in itself. The value of ends is intrinsic, and 'the value of the means derives from the value of the end, and is, moreover, secondarily determined by how efficiently it serves it'.[9] Grant is thinking of technology

[8]   Aristotle, (D. Ross, trans.), *Nichomachean Ethics*. (Oxford: Oxford University Press, 1998).

[9]   Robert Grant. 'Values, Means and Ends', in Roger Fellows, Ed., *Philosophy and Technology*. (Cambridge: Cambridge University Press, 1995), 183.

James Garvey

generally when he concludes that 'the primary value of technology is instrumental, and is determined by its ultimate end'.[10] For Grant, the ultimate end, in so far as technology generally conceived has an ultimate end, is freedom from drudgery, a kind of liberation.

If our interest is the moral evaluation of particular instances of technology, Grant goes on to argue, we can note not just the freedom the object gives us, but also the ends we have in view when we use it. Technological objects themselves cannot be rationally criticized except with respect to their adequacy as means to certain ends. It is making the right choices with regard to ends that is the entire subject-matter of the moral evaluation of technology. 'It is precisely because technological processes and procedures are neutral ...that they, or more accurately the ends they subserve, require moral scrutiny'.[11]

Is it true that means are neutral, acquiring moral value only derivatively as means to intrinsically valuable ends? I get the feeling that there is more to it than this. Suppose you and I want to do something about poverty in London. We think we can do the most good by helping poor children, so our end in view is lowering the rate of child poverty in London, which stands at, say, 10% of the child population. We embark on means-ends reasoning, and I offer you the following suggestion: we can lower the level of child poverty by distributing poisoned sweets in the schools of Southwark. We've got a good end in view and an efficient means which serves it. It won't cost much, and the percentage of poor children in London really will go down very quickly if we follow my plan. If means are neutral, acquiring value only derivatively with reference to an end, and our end is a good one, then what's the problem with the means I'm contemplating?

You might want to object and say that the end we have in mind has not been carefully articulated. What we meant was not just a change in the numbers, but eliminating poverty or some other, more clearly good good. Have a think about all the usual objections to utilitarianism which have to do with justice, scapegoats, and moral intuitions concerning the dubious means a good utilitarian might be forced to countenance in order to bring about a good end, namely the greatest happiness for the greatest number. The problem for the utilitarian and the instrumentalist view of technology is that means are not obviously neutral, and when we evaluate them we have more than just efficiency in mind.

[10] *Ibid.*, 181.
[11] *Ibid.*

The independent reasons you might have for finding means dubious seem to be of two sorts. First, you might have a problem with the means themselves. Second, the means might secure a desired end, but they do more besides, and you might be worried about the other consequences the means bring about. Think again about distributing poisoned sweets in Southwark. Someone might object to the implementation of these means by saying, 'But that's murdering children, and murdering children is wrong.' or by saying, 'That will lower the poverty rate, but it will also result in dead children, and that consequence is wrong.' Something might hang on how finely we describe actions, but I'm going to ignore it for now. In the first reply, the problem is the means themselves. In the second, there is still a sliver of the view that means acquire value from ends, but it's no longer just the ends we have in mind which matter. Means sometimes have consequences in addition to those intended. So there has to be something wrong with thinking that the value of means comes only from reflection on the ends towards which they are directed.

Well, all right, you might think, but the question is about the neutrality of technology and you have been going on about poisoning children. The examples do not match up. In one case we are talking about means as objects used in actions and in the other case we are talking about actions themselves. Actions can be morally evaluated by considering the actions or their consequences. Objects are just objects. Of course, we do things with technology, and when we do something with technology we can evaluate our actions in the usual way, but the objects have no consequences. We can use them in various ways, and it's the various ways we use them which gives them a moral flavour. The objects are still morally neutral.

We have slipped rather easily from talking about means and ends to talking about the functional promiscuity of technology, what might be part of a different argument for instrumentalism. It might be claimed that technological objects can be put to many different uses. The objects are not for anything in particular, and this is where the thought that objects are morally neutral originates. A hammer might be used to build a house or as a murder weapon. Technological objects, like stem cells, can turn into anything. Precisely because of the functional promiscuity of technology, technology considered just in itself is neutral.

A number of thoughts could and probably should be disentangled here, but I'll focus just on the presupposition that objects are neutral because they are not for anything in particular. It is this

James Garvey

claim or something like it that the argument from functional promiscuity needs for the conclusion that objects are neutral. You can hear the connection between promiscuity and neutrality in one of Grant's claims, a passage we already noted but one worth considering in full in this connection: 'technological processes and procedures are neutral—that is to say, employable in principle in any cause, good or bad ...'[12] No one disputes the claim that objects can be used in different ways, but does this force us to accept the further claim that objects are not for anything in particular?

Promiscuity and neutrality are not tied together as strongly as might be expected. I know that a gun can be used as a paperweight or a hammer, but that need not stop me from joining Kurt Vonnegut in thinking a gun is 'a tool whose only purpose was to make holes in human beings'.[13] The two views are not inconsistent. You can think of an object as having a primary function or functions, as well as other possible functions.

Noticing the consistency of the claims is enough for me to reject the alleged tie between neutrality and promiscuity, but you might want more to go on. There are two lines of thought here, both issuing in the claim that objects are for something in particular, and both are a little shaky. According to the first, we cannot help but see objects as for something; according to the second, objects really are for something, whether we see it or not.

One might say that although it is possible, strictly speaking, to abstract an object away from its context and thus come to think of it as neutral 'in itself', objects are never really presented to us in this way. We are social creatures and language users, and as such we know what things are for. You might, in other words, argue against instrumentalist neutrality by claiming that objects suggest a use to us, given our social or cultural background. No one growing up in the West can see a gun and think of all its possible uses as on a par with its primary use. This might be the line Heidegger takes when he says that an object might be seen as 'present to hand' or 'ready to hand'.[14] Tools, certainly, fall into the latter category.

You might be persuaded that human beings see tools as for something because our brains have evolved to do just that. Tools are so much a part of our evolutionary history that we have evolved

[12] Grant, 187.
[13] Kurt Vonnegut, *Breakfast of Champions*, (London: Cox & Wyman, 1992), 49.
[14] Martin Heidegger, (John Macquarrie and Edward Robinson, trans.) *Being and Time*, (New York: Harper and Row, 1962).

250

to pick them out as quickly as we notice potential mates or enemies. I don't want to make very much of this, but there might well be empirical support for the view, coming from certain cases of brain injury or lesion. Individuals with damage to the posterior inferior temporal lobe and anterior lateral occipital region are entirely unable to name tools, and tools only.[15] Some might be willing to conclude from this that normally functioning human brains have modules dedicated to seeing objects as for something, seeing objects as tools.

The foregoing reflections might nudge you towards accepting the view that a piece of technology is for something despite its functional promiscuity, but you might still have one last worry. Even if we see objects as being for something in particular, that does not mean they are for something in particular. Considered just in themselves, abstracted away from social contexts, objects really are neutral. I confess that I am having some trouble thinking about something like a hammer in this sort of abstraction. And, anyway, even if this sort of thought is possible, do you not end up with the view that anything you like is neutral when abstracted away from the relevant contexts—a piece of fine art or a selfless act?

Put the question aside, because we need not answer it. If you are after the claim that objects have value in themselves, there are some considerations which might be suggestive. Objects are, after all, designed by people with particular uses in mind. As Monsma argues, technology is value-laden because technological objects are unique creations, designed to function in a particular and limited way.[16] The point is put a bit more strongly by Franklin, who might not have read Hume, and who goes so far as to say that an object's use is 'incorporated a priori in the design and is not negotiable'.[17] You need not go quite so far to conclude that because human beings build things for certain purposes, those things are suited to some activities and not others. The things themselves might well have inbuilt, primary functions.

I am not sure what to think about all of this. What is clear, though, is the fact that objects might be put to many uses does not

[15]   Martin, A, J. V. Haxby, F. M. Lalonde, C. L. Wiggs, and L. G. Ungerleider, 'Discrete Cortical Regions Associated with Knowledge of Colour and Knowledge of Action', *Science* **270**, 102—105.
[16]   S. U. Monsma (Ed.) *Responsible Technology,* (Grand Rapids, MI: Eerdmans, 1986).
[17]   Ursula Franklin, *The Real World of Technology*, (Toronto: House of Anansi Press, 1999), 18.

undercut the conviction that objects have a primary or set of primary uses. You can think that objects suggest a use to us because of our culture, or because the objects are built to fulfil some purposes and not others, or, I suppose, because objects are in fact used by us primarily in some ways and not others. However you choose to think about it, you are left with the view that objects are for something. They really do have particular consequences. They are not neutral in the required sense.

Is the instrumental view much help in thinking about the moral dimension of mundane technology? We have noticed three mistakes in it. First, it is a mistake to have our eyes only on Baconian delights in reflection on the moral value of technology. Second, it is a mistake to think that we evaluate technology just in terms of the ends we intend. Sometimes we evaluate the means themselves and sometimes we evaluate the means with an eye on consequences other than those intended. Third, it is a mistake to make too much of the fact that objects might be used in different ways. Objects are for some purposes and not others, and this can give us reason to suspect that they are not neutral. If the neutrality view is a non-starter, we can look elsewhere for help in thinking about mundane technology, namely, we can consider the autonomy view.

2. The autonomy view

You have only to dip your toe into the literature on autonomy, which is about as much as I have managed, to realise that it is a swamp. Finding a common claim or set of arguments is not easy. Certainly there is a rejection of the neutrality of technology, but the rejection is based on something more like a Gestalt shift than an argument. The main thesis, if there is just one, is put in a number of ways, but the general idea is that technology is somehow autonomous, not under our control. Winner, for example, identifies several expressions of the thesis:

> In some views the perception of technology-out-of-control is associated with a process of change in which the human world is progressively transformed and incorporated by an expanding scientific technology. In others the perception focuses upon the behaviour of large-scale technical systems that appear to operate and grow through a process of self-generation beyond human

intervention. In others still, the matter is primarily that of individuals dwarfed by the complex apparatus surrounding them ...[18]

There are three themes in here, and we'll briefly consider each of them.

Talk of a general change in the nature of 'the human world' shows up in many places. Feenburg, for example, claims that 'the issue is not that machines have "taken over" but that in choosing to use them we make many unwitting cultural choices. Technology is not simply a means but has become an environment and a way of life ...'[19] The first part of the claim, choosing to use technology involves us in unwitting cultural choices, sounds familiar. It sounds like an enlarged version of something noticed in our consideration of neutrality: namely, technology sometimes brings with it more ends than those intended, and unintended ends sometimes figure into our evaluation of means. Certainly the autonomy view holds that technology has more effects than the ones we intend, but the view expands this simple claim into something much larger. The unintended effects are now part of a whole system, a thing with identifiable and in some sense self-perpetuating ends which lie beyond our control.

Probably this notion of technology as a system—which emerges in the second part of Feenburg's thesis, the claim that technology is an environment or way of life—falls out of Heidegger's interpretation of technology.[20] Heidegger's question concerning technology has something to do with our relationship to technology, in particular, our experience of being 'in the midst' of a life surrounded by or embedded in technology. He claims that the view of technology as nothing more than a means might be correct as far as it goes, but it points to a deeper understanding of the relationship between human beings and technology. We see technology as tools to be used and we see problems as admitting of technological solutions precisely because we are already embedded in something deeper and larger, something like a technological way of life.

Saying that we live a technological way of life is claiming that our problems, activities, agendas, and so on happen in a social world

[18]   Langdon Winner, *Autonomous Technology*, (Boston: MIT Press, 1977), 17.
[19]   Feenburg, 8.
[20]   Martin Heidegger, 'The Question Concerning Technology', *Basic Writings*. (New York: Harper Collins, 1993), 331—41.

James Garvey

where everything is in some measure regarded as a standing-reserve, a stockpile of stuff to be used in the service of technological purposes. We, along with everything else around us, are nothing more than instrumentally useful, disposable human resources, in a way of life we never chose.

Winner's second expression of the autonomy view—the one which focuses on the behaviour of large-scale technical systems that appear to grow through a process of self-generation beyond human intervention—is taken literally or nearly literally by some thinkers. Technology is not just a force, but a creature with its own agenda. Ellul, perhaps the loudest spokesperson for the autonomy thesis, maintains that, 'An autonomous technology ...means that technology ultimately depends only on itself, it has its own route ...it must be regarded as an "organism" tending toward closure and self-determination: it is an end in itself.'[21]

Autonomy, for Ellul, consists partly in technology's detachment from things we might think regulate it. In fact, the sorts of things we might think are in charge of technology are actually beholden to it. Technology operates outside of morality and aims at self-perpetuation rather than improving human virtue. It is also not something researchers or users tend to judge from a moral point of view. Technology answers technological problems, not moral ones. Because technology seems to operate outside of morality, he argues that we never halt research or applications for moral reasons—instead, technological necessity drives implementation and use. This is not to say that technology is morally neutral, but instead the claim is something closer to the view that 'technologically good' replaces 'morally good'—what is good for the system overrides what is good for us.

Interestingly, the claim that technology is not just beyond moral considerations but that it is now an arbiter of moral problems is considered in this connection. Technology, Ellul suggests, is so powerful a force that moral problems get technological answers or, worse, some moral considerations are put to one side as technologically impractical. Similar considerations hold for economic, political and scientific constraints on technology. Technological demands drive everything. The thesis which emerges from Ellul's reflections is technological determinism, the view that changes in technology dictate or fix social and cultural changes.

---

[21]   Jacques Ellul, (Joachim Neugroschel, trans.) 'Autonomy', *The Technological System*, (New York: Continuum Publishing, 1980), 125, but see his footnote one.

Winner's third view of autonomy—a conception of technology as a complex which dwarfs the individual—gets only occasional attention. Winner himself argues that there is a substantial gap 'between complex phenomena that are part of our everyday experience and the ability to make such phenomena intelligible and coherent.'[22] Expertise and know-how necessarily fragment, fall into the hands of a clutch of specialists in a technological society. No one understands even a fraction of the science underpinning even everyday objects, and it almost goes without saying that the interconnections between various objects and their social effects are generally beyond our ken. It can lead to a kind of vertigo. Worse, it can result in a general degradation of individual autonomy, which, some argue, is replaced by the autonomous activities of technology itself.

Cooper does not go so far, but he does identify three senses in which technology erodes the sphere of the person.[23] Self-reliance is diminished by technology, in so far as technical or expert fixes are sought, rather than individual reflection on a problem. Not only is there now a culture of expertise in which individuals defer to the opinions of experts, but in some sense this is a necessity, given the complexity of technology. Further, he argues that the integrity of our concept of personhood is under a kind of threat from the way in which technology reinforces distinctions between certain human capacities and elevates some above others. Instead of reason, intuition, emotion, wisdom and knowledge, we worry about acquiring and processing information. Finally, the aspect of self which Cooper characterises as a 'life-as-a-whole' requires a certain narrative structure undercut by the speed of technological advance. Our concept of self requires not just past memories, but forward planning, hopes for the future. Cooper argues that living in an age of rapid technological change undercuts the possibility of such plans. Our conception of ourselves requires slow plodding and is impaired by the speed of technological progress.

You have now heard something about the thesis or theses of autonomy, but what arguments are there for the view? As noticed a moment ago, what we have is something closer to a Gestalt shift than an argument. If there is an argument somewhere in here, it consists largely in the claim that a wide-angled shot of technology is the only way to see it properly and come to understand it. Recall

[22] Winner, 282.
[23] D. E. Cooper, 'Technology: Liberation or Enslavement?' in Fellows, 1995.

James Garvey

Kline's claim that technology is best thought of as sociotechnical systems of manufacture and use. His claim is that technology cannot be understood for the thing that it is without reference to the whole or at least large chunks of the society in which it is embedded, including the purposes and aims of human beings. Building all this in, he says, is 'essential to understanding the human implications of "technology" '.[24] If you want something more than a naïve and limited conception of technology, the autonomy theorist seems to argue, you need to have a look at the larger picture. Many thinkers in the neighbourhood take it that the broad view requires a conception of technology as a system or force with its own implications and agenda. Examples are used to nudge us towards the view, and considering some of them might help.

Winner argues that technology might be thought of as autonomous in at least two senses. First, 'some kinds of technology require their social environments to be structured in a particular way in much the same sense that an automobile requires wheels in order to move.'[25] His claim is, roughly, that some sorts of technology function best or maybe function only against the backdrop of some social or political arrangements and not others. Using a particular technology, then, can either reinforce or require a particular social setup. His example here is nuclear power facilities. Keeping such things safe and coordinating the complexity underpinning the production of nuclear power requires centralised control. If you want nuclear power, you need to have and then keep a particular political arrangement. He concludes that just having nuclear power facilities in a society involves that society in a measure of authoritarianism.

Second, he argues that the invention, design, and arrangement of a particular technical device or system settle issues in the affairs of a given community. His example here concerns the bridges around Jones Beach, and his conclusion is that they literally have an inbuilt political or social effects. Once implemented, the system is beyond the control of its users. It carries on doing what it does regardless of our desires or aims, and what it does is political. The bridges have politics.

The point of these and other examples which are used by autonomy theorists is that unless some version of the autonomy thesis is adopted, unless we opt for the bigger picture, we won't

[24] Kline, 217.
[25] Langdon Winner, *The Whale and the Reactor*, (Chicago: University of Chicago Press, 1986).

notice the larger social, political and moral implications of technology. You can take this point, but still wonder whether or not one or another version of the autonomy thesis is forced on us. Do we really have to see technology as a way of life, or an independent organism, or a complex of forces which somehow overwhelms the self in order to make sense of these sorts of examples?

The conclusion that technology is out of control, however you cash it out, can seem extravagant if all you have to go on are suggestive examples. I'll leave this question to you, because the point of considering the autonomy view, at present, is to determine whether or not it can help in thinking about the moral dimension of mundane technology. Suppose it's true that technology is beyond our control: what should we do about it? Is there any help here for someone worried about the moral use of everyday technology? Is there any good advice in the autonomy view?

We might look to Heidegger for advice about what to do if technology really is a system, a life world in which we are embedded. His suggestion seems to be that we should 'thoughtfully reflect' on the senses in which we are enveloped by technology, instead of further attempts to save ourselves with still more technological quick fixes. We can look to art, he seems to say, and build something like an aesthetic point of view into our relationship to the technological world. The suggestion is that we should think carefully about our world, perhaps think of the mountain as beautiful rather than as a possible source of coal. Very well, but you might be forgiven for thinking that we need a bit more to go on than this.

What about the view that technology is an autonomous creature or something like it? If you join Ellul in thinking that technology determines and drives more or less everything, it is hard to think of a way to snap ourselves free of it. Determinisms do not normally lend themselves to clear thinking about possible courses of action. Still, Ellul is careful to say that his analysis of technology and his conclusions about technological determinism leave open some sort of human freedom—human beings need not just float along with the technological tide. Ellul's purpose is, he says, 'to arouse the reader to an awareness of technological necessity and what it means. It is a call to the sleeper to awaken'.[26] He has, though, 'deliberately refrained from offering solutions'. The best advice he

[26] Jacques Ellul, (John Wilkinson, trans.) *The Technological Society*, (New York: Jonathan Cape, 1964), xxxiii ff.

has for us is that 'each of us, in his own life, must seek ways of resisting and transcending technological determinants'. Again, not much help.

Think again about the third version of the autonomy thesis, the claim that human agency is somehow swamped by technological complexity. There are some calls in this connection for a simple reduction of complexity, of insisting on small-scale technological solutions to problems, the so-called 'appropriate technology movement.' The trouble with this and other suggestions for a reduction in complexity is that we have no idea where to begin. Technology is already in place, and the complexity of technological systems is such that we have no real idea what scaling back or simplifying particular parts of the whole will do. Winner is alive to this worry, and suggests that we adopt Luddism as epistemology.

After offering a few general rules for the future implementation of technology, Winner suggests that we dismantle some technological systems, not as a solution to the troubles associated with the autonomy of technology, but as a method of inquiry. 'Prominent structures of apparatus, technique and organization would be, temporarily at least, disconnected and made unworkable in order to provide the opportunity to learn what they are doing for or to mankind'[27] Once we know what we are doing, or better, what technology is actually doing, we might use that knowledge to reconfigure the technological world, direct it more carefully towards more human ends. If this sort of thing is thought to be impracticable, Winner suggests that we experiment with a group of willing individuals, extract them from technological systems, and see what this tells us about the role of technology on social relationships. We might also, as a people, just choose not to repair technological systems when they happen to break down and examine the consequences.

I am not entirely sure that Winner is serious about any of this. Talking about Luddism as epistemology might be his round about way of pointing to probably the largest problem attending reflection on technological complexity: we do not really know enough about the effects of technology to make reasonable suggestions for improving the situation. If Winner is serious, then there is nothing here which counts as advice. We can only await the results of the experiments, and we might be waiting for some time.

So much for the autonomy view. We have reflected on versions of the autonomy thesis, considered the examples which go along with

[27] Winner, *Autonomous Technology*.

the recommendation that technology should be viewed as a system out of control, and looked for help with our question about mundane technology. In the end, even if you do go along with the autonomy thesis, the view isn't much help, doesn't offer much in the way of advice.

3. A Concluding Suggestion

Philosophers who take up questions associated with exciting technology usually have something to say about how it is best used, and typically there is a call for limiting its use or at least deferring the question of its use until wisdom catches up with science. You might think that demanding advice from the neutrality view and the autonomy view is misplaced, but if we get recommendations from nearby theorists, why not hope for a little practical help with regard to mundane technology? Can we tease anything out of our considerations of both views?

If, despite the dubious arguments, you go in for some version of the neutrality thesis, you end up with just one thing to think about: the ends towards which technology is directed. If you are persuaded that there is more to it than this, you might begin to go further by reflecting on the risks involved in using technology, engage in a bit of cost-benefit analysis. Technology generally, Grant for example notices, brings with it the risk of temporary unemployment, the depletion of resources, climate change and other unpleasant effects. A part of the autonomy theorists' objection to even this short step beyond simple neutrality is that reflection on just the ends in view and the risks is not enough. More consequences than just the intended ends are brought about by technological use, and the concept of risk is not meaty enough to capture those consequences. Thinking about risks also implies that we are in control of the situation, but of course the autonomy theorist denies this.[28] Must we be dragged off to the view that technology is autonomous if we want to think clearly about those extra consequences?

If you go in for some version of the autonomy thesis, you find no serious advice at all. The wide-angled view might capture the large scale social, political and moral implications of technology, but it takes the human perspective right out of our thinking. The

[28]   Langdon Winner, *The Whale and the Reactor*, Chapter 8.

autonomy theorist might be forgiven for this. If technology really is beyond our control, then what advice could there possibly be?

There ought to be middle ground in here somewhere, some conception of technology which recognises that the stuff is more than just neutral but stops short of contending that technology is beyond our control. We might do better by thinking of technology as sometimes out of control, not beyond control, and do what we can to change the undesirable facts in our everyday lives. If what is needed is a more human perspective, you can look for one by turning Winner's possibly tongue-in-cheek advocacy of Luddism as epistemology into a real suggestion, something like personal Luddism.

Unplug your television for a week or two, go on holiday without your camera, consign your Ipod to a drawer, ignore your email, leave the phone off the hook for a while, and study the effects on your life. Perhaps you switched off your mobile phone for a bit of peace while reading this paper. You could leave it off for a while, or at least make sure you have a good reason for turning it on again. It might well be that reflection on your good reason when it comes, as well as reflection on reasons for leaving the thing off a little while longer, will lead to clear thinking about the moral status of mundane technology. You might find advice for yourself. This strategy can seem more fruitful than the further consideration of the theses of autonomy and neutrality.

Is Simplicity Evidence of Truth?[1]

ADOLF GRÜNBAUM

In a short 1997 book entitled *Simplicity as Evidence of Truth*, the Oxford philosopher Richard Swinburne has put forward the following thesis summarily: '... for theories (of equal scope) rendering equally probable our observational data (which, for brevity I shall call equally good at "predicting"), fitting equally well with background knowledge, the simplest is most probably true'.[2]

It is crucial that the explanatory theories which Swinburne is concerned to compare as to simplicity are *competing*, incompatible or rival theories.[3] As he put it:

> But what are the criteria for supposing that some purported explanation, some one explanatory hypothesis *rather than some other one* provides the true explanation? There are, I am going to suggest, two a posteriori and two a priori criteria for determining this. These criteria have to be fairly well satisfied as a whole in order for us to be justified in supposing that we have an explanation at all. But, given that, then *among **incompatible purported** explanations* [theories] the one which satisfies the criteria best on the whole is the one which is most probably true' [italics and bolding added].[4]

Furthermore, as shown by the quotation from Swinburne in our opening paragraph above, it is of critical importance that these conflicting theories are required to be *of equal scope* at the very outset, although he oddly enough puts that requirement into parentheses. Indeed, the problem of warranting that two rival theories have the same *scope* will turn out to be the Achilles heel of his prescription for identifying one of them as simpler than the other, and as more likely to be true.

---

[1] This paper was also delivered at All Souls College, Oxford University (March, 2006).

[2] R. Swinburne, 'Reply to Grünbaum', *British Journal for the Philosophy of Science* **51**, No. 3, (2000), 481.

[3] R. Swinburne, *Simplicity as Evidence of Truth* (Milwaukee, WI: Marquette University Press, 1997), 11–12.

[4] Ibid., 11–12.

Adolf Grünbaum

Swinburne's short 1997 book was based on his Aquinas Lecture of that year at Marquette University (USA). More recently, in 2001, he published a book on *Epistemic Justification* in whose Chapters 3 and 4 he merely amplified his 1997 Lecture and gave it a fuller defense.

In his 2001 account, he does not give any characterization of his notion of the 'scope' of a theory,[5] but in his Aquinas Lecture,[6] he had recourse to the received cognate notion of the logical 'content' of a theory. The logician Alfred Tarski introduced the concept of 'the consequence class of a statement S' as the set of *all* the deductive consequences of S. This consequence class has also been denominated as the logical *'content'* $LC(S)$ of S.

If T_1 and T_2 are incompatible theories, then the logical content of T_1 is *not* a subset of the content of T_2, nor conversely. The reason is that each of their logical contents has at least one member that does not belong to the other, which makes their contents *incomparable* with respect to set inclusion. But, as we saw, Swinburne's prescription for meaningful simplicity comparisons of rival theories is crucially predicated on the *equality* or sameness of the respective contents or scopes of these theories.

How then does he effect comparisons of the contents of clashing theories with a view to showing that their contents are indeed equal, as required for rating these theories in regard to comparative simplicity? Alas, Swinburne's answer is a largely elusive *ipse dixit* as follows: 'There is no precise way of measuring content, but we can compare content. If one hypothesis entails another but is not entailed by it, the former has more content than the latter'.[7] And to this very rudimentary statement, he added more recently: 'And rough comparisons [of content] at any rate, and sometimes precise comparisons, are possible between theories not thus logically related [by one-way entailment]'![8]

But note that his example of content-comparability between theories one of which unilaterally entails the other is quite beside the point in this context; instead, the issue is that Swinburne needs content-comparability between *incompatible* theories. By the same token, even assuming that content-comparisons were possible between *some* theories neither of which unilaterally entails the

---

[5] R. Swinburne, *Epistemic Justification* (Oxford: Clarendon Press, 2001), 82.

[6] Op. cit. note 3, 12–13.

[7] Op. cit. note 3, 13.

[8] Op. cit. note 5, 82.

other, as claimed by Swinburne, this comparability does not vouchsafe the content-comparability of *conflicting* theories, such as Newton's and Einstein's, which are avowedly indispensable for Swinburne's simplicity ratings.

Thus, as we shall see, it is a cardinal and indeed fatal failure on his part that he does nothing to redeem his crucial claim that at least 'rough comparisons [of content] are possible' between theories not related by a relation of unilateral entailment. But Swinburne urgently needs content-comparisons between *competing* theories to effect his simplicity ratings, and yet theories unrelated by unilateral entailment may well not be *rival* theories.

Swinburne elaborates on criteria other than simplicity, such as empirical adequacy, that may warrant one theory to be 'more likely to be true' than another. For example, as when 'one theory is superior to another in yielding the data to a higher degree of inductive probability, or in yielding more data to the same degree of probability'.[9] And his elaboration features the following major disclaimer:

> Having made these points about the other principles at work [besides simplicity] in assessing the probability of scientific theories, I emphasize that I have no theory to put forward about how the factors involved in them are to be measured, nor about how in comparison between two theories better satisfaction of one criterion is to be weighed against worse satisfaction of another. Rather, I aim to show the nature and force of the criterion of simplicity by considering cases of *conflicting* theories where the other criteria for choice between them [such as parity of *content*] *are equally well satisfied* [italics added].[10]

In short, as Swinburne tells us concisely, he wishes to deploy the criterion of simplicity 'in choosing among [competing] scientific theories of equal scope [i.e., content] fitting equally well with background evidence and yielding the same data with the same degree of logical probability'.[11]

But since the decisively relevant content-equality for Swinburne's simplicity ratings is between *rival* theories, it was unhelpful on his part to offer the following illustration of *compatible* theories of supposedly equal content: 'A hypothesis which predicts all the positions of Mars for the next century would have the same content

---

[9]   Op. cit. note 2, 13.
[10]   Op. cit. note 2, 14–15.
[11]   Op. cit. note 4, 83.

Adolf Grünbaum

as one which predicted all the positions of Venus for the same century'.[12] Unfortunately, Swinburne offers no *representative* illustration at all of his thesis of at least rough content-comparability of *conflicting* theories, let alone a systematic treatment of such comparability.

It is indeed regrettable that he did not heed the highly relevant sobering lessons spelled by the multiple failures of Karl Popper's attempts to use content-*increase* as between rival scientific theories to characterize scientific progress. I have given a detailed demonstration of that highly instructive Popperian debacle in a series of my papers in the 1976 volume of the *British Journal for the Philosophy of Science*.[13], [14], [15]

To articulate the range and depth of the difficulties besetting Swinburne's imperative need to warrant content-equality for conflicting theories, it behooves me to recapitulate my demonstration of how Popper failed fundamentally in rank-ordering incompatible theories with respect to content-increase.

He has adduced Einstein's and Newton's rival theories of gravitation as a poignant instance of the following claim made by him: If a non-metrical notion of the content of a theory T is construed as the set of all those *questions* to which T can provide *answers*, then the contents of at least some logically *incompatible* extant scientific theories are qualitatively comparable with respect to the relation of proper inclusion. Let me denote the set of questions that can be answered by a theory T as its question-content, otherwise known as 'erotetic' content, $QC(T)$. And let us use the term 'Thesis Q' to refer to Popper's claim that the question-content $QC(N)$ of Newton's theory is a proper subset of the question-content $QC(E)$ of Einstein's theory as follows:

(a) to every question to which Newton's theory has an answer, Einstein's theory has an answer which is at least as precise; this makes (the measure of) the content, in a slightly wider sense than Tarski's, of N less than or equal to that of E; (b) there are

[12] Op. cit. note 4, 82.

[13] A. Grünbaum, 'Can a Theory Answer more Questions than one of its Rivals?' *British Journal for the Philosophy of Science* 27, No. 1, 1976, 1–23.

[14] A. Grünbaum, 'Is the Method of Bold Conjectures and Attempted Refutations Justifiably the Method of Science?' *British Journal for the Philosophy of Science* 27, No. 2, 1976, 105–136.

[15] A. Grünbaum, 'Ad Hoc Auxiliary Hypotheses and Falsificationism', *British Journal for the Philosophy of Science* 27, No. 4, 1976, 329–362.

questions to which Einstein's theory E can give a (non-tautological) answer while Newton's theory N does not; this makes the content of N definitely smaller than that of E.

Thus we can compare intuitively the contents of these two theories, and Einstein's has the greater content.[16]

Popper tempers his claim of the erotetic comparability of rival theories such as Newton's and Einstein's by noting that 'there are also competing theories which are not comparable'.[17] We shall designate Popper's two conditions in his statement of his Thesis Q as requirements (a) and (b) respectively.

But, in my essay 'Can a Theory Answer More Questions than one of its Rivals?'[18] I have shown that Popper's Thesis Q is untenable, because there are well-formed questions answerable in Newton's theory that are not answerable in Einstein's. Here, I shall provide several illustrations of this erotetic discrepancy. Indeed, Newton's and Einstein's theories provide a poignant illustration of the incomparability of the erotetic contents of competing theories which feature basic conceptual disparities and yet are both addressed to a given domain of phenomena.

Most importantly, as we shall see, the failure of Popper's Thesis Q and of his content-increase criterion for the admissibility of auxiliary hypotheses will spell a dismal prospect for whether Swinburne could ever redeem his aforecited claim that 'rough comparisons [of content] at any rate, and sometimes precise comparisons, are possible between theories' not related by entailment. Astonishingly, this pivotal claim is altogether Pollyannish, a mere *ipse dixit*.

Let us consider several sorts of counterexamples to Popper's Thesis Q that the question-content $QC(N)$ of Newton's theory (N) is a proper subset of the question-content $QC(E)$ of Einstein's theory E.

In the context of the general theory of relativity (GTR), the special theory of relativity (STR) is the theory of Minkowski flat space-time and is thus a special case of the GTR, which is Einstein's theory of gravitation. Popper does not distinguish

[16] K. R. Popper, *Objective Knowledge* (Oxford: Clarendon Press, 1972), 52–53.

[17] Ibid., 52.

[18] A. Grünbaum, 'Can a Theory Answer more Questions than one of its Rivals?' *British Journal for the Philosophy of Science* **27**, No. 1, 1976, 1–23.

Adolf Grünbaum

between them. Now, among bona fide Newtonian questions that will turn out to be unanswerable in Einstein's theory, some of them will be demonstrably unanswerable in the STR, while others are unanswerable in the GTR, or are unanswerable in both.

Case 1. In Newton's dynamics, the velocity v attained by a particle of fixed mass m which is accelerated from an initial state of rest by a constant force F acting for a time t is given by

$$Ft = mv.$$

Suppose that one were to ask N for an answer to the question:

(1a) What *finite* amount of time is required by a constant force of magnitude F to accelerate a particle of fixed mass m in an inertial frame up to a speed $2c$, which is twice the numerical value of the usual speed c of light *in vacuo*?

Clearly, N would answer precisely that $t = 2mc/F$. But how would E deal with such a question? The STR asserts that the fixity of m when the particle is no longer at rest is a false presupposition of the question (1a), as is the latter's presupposition that any superlight velocity such as $2c$ can be attained at all by a particle. Thus in E, the demand of (1a) for a time specification t would be ill-conceived, even if t is allowed to go to infinity! How then can E (STR in this case) possibly be held to answer (1a) at all, let alone at least as precisely as N does?

By the same token, there are infinitely many questions about the outcomes of, say, collisions of positive mass particles under specified initial conditions involving their having superlight velocities in (local) inertial frames, such that these questions can be *asked* and *answered* in N but not in E, let alone with at least equal precision. For E denies the very physical possibility of the initial conditions on which each one of this infinitude of questions is predicated, whereas N asserts these same initial conditions to be physically realizable. Even just this class of cases involving Newtonian particles traveling at superlight velocities suffices to impugn Popper's Thesis Q as follows:

In the case of empirically competing (incompatible) and moreover conceptually disparate theories having different ontologies, any question asked in either theory which is predicated on an assumption denied by the other, or not even *statable* in the other's conceptual framework, will not be answered by the other (with at least equal precision), but at best will be *obviated* by the other.

Case 2. Consider the following dynamical example. As before, let us denote Newton's theory by N, and Einstein's by E. In Einstein's theory of gravitation, the orbit of a planet of negligibly small mass *subject solely to the sun's gravitational field* is *not* a perfectly closed ellipse about the sun as it is in Newton's theory. Instead, the orbit of such a planet is a slowly rotating ellipse.[19] Now let us ask the question:

(2a) Why is the orbit of a planet of negligible mass which is subject solely to the sun's field a perfectly closed ellipse about the sun?

Clearly this question *is* answered in N in the proper sense of exhibiting the presupposition of the question to be a consequence of N's theory of gravitation. In what sense, if any, does E 'answer' question (2a)? In the context of E, (2a) is predicated on a false presupposition and is thus an abortive non-starter. Can E therefore be *absolved* from having to *deal* with (2a) at all? And if so, can E be held to have 'answered' (2a) merely because E may be absolved from having to *deal* with it at all?

It is true enough that it is *not* automatically incumbent upon a theory B to provide an answer to a question Q merely because Q is well-conceived and answered in a *rival theory* A in which some presupposition p of Q is held to be correct: If p is held to be either gratuitous or false in B but is asserted by A, then B's failure to answer Q in the sense of providing a *reason* for p is not necessarily a defect of B. By way of analogy, suppose that a court either has no information showing that a certain witness ever beat his wife at all or has information that the witness has never beaten his wife. Then it is true enough that the judge cannot regard the witness as uncooperative when the witness fails to supply a date in answer to the question 'when did you last beat your wife'?

But how does this help Popper make good on the claim that question (2a) *is* 'answered' in E at least as precisely as it is in N, a claim which is essential to the fulfillment of requirement (a) of his Thesis Q?

Since E rejects question (2a) as being predicated on a false presupposition and hence as ill-conceived, E can be said to *obviate* this question, rather than to answer it. Thus, in the context of the requirements of Thesis Q, such obviation of a given question by a particular theory does not count as an answer to the question. Indeed, the need to obviate questions that are unanswerable in one

---

[19] A. Grünbaum, *Philosophical Problems of Space and Time* (Dordrecht: D. Reidel, 2d ed. 1973), 577, n. 11.

of two incompatible theories N and E, while being answerable in the other, is an expression of the clash of some of their presuppositions.

So much for the failure of Popper's Thesis Q, which spells a very sobering challenge indeed to Swinburne's aforecited entirely unsubstantiated claim that: '... rough comparisons [of content] at any rate, and sometimes precise comparisons, are possible between theories'[20] not related by one-way entailment.

It would seem that E and N are hardly good candidates for the content-equality required by Swinburne to compare them as to simplicity. But they likewise seem to be poor specimens for overall comparative ratings of *simplicity*. As Swinburne appreciates, when considering the relative simplicity of a scientific theory, it is crucial to specify the particular respect(s) in which one theory is held to be simpler than another. After all, theory B might be simpler than theory A in *one* respect, while being more complicated in another. Thus, an *explanatory unification* of previously disparate sorts of theoretical elements of a prior theory, when achieved by a *new* theory, could be taken as a mark of its greater simplicity vis à vis its predecessor, even though the two theories might exhibit an *inverse* simplicity-ordering in some *other* respect. In this vein, the GTR effected a *simplifying unification* of the description of the physical geometry with a theory of gravitation, which had been quite distinct in its Newtonian predecessor. Thereupon, the GTR absorbed gravitation in the metric tensor g_{ik} of its space-time geometry.

But observe that this simplifying unification does *not* owe its *warrant* at all to some greater *a priori* simplicity; instead it derives its justification from such empirical promptings as the equivalence of gravitational and inertial mass, which Eötvös had demonstrated experimentally. So much for *one* respect in which the GTR is simpler than Newton's theory of gravitation.

Yet there is another respect in which Newton's theory of gravitation is indeed simpler than Einstein's. The awesomely complex non-linear partial differential equations in Einstein's field equations are far less simple than the ordinary second order differential equations in Newton's law of universal gravitation. Swinburne requires comparative ratings of *overall* simplicity for his program of probability assignments to competing theories of equal content. But at least the case of Newton's and Einstein's theories may indicate that his notion is a will-ó-the-wisp.

[20] Op. cit. note 4, 82.

Case 3. As I have shown on an earlier occasion,[21] a highly instructive case of content-incomparability is furnished by Popper's unsuccessful attempt to legitimate theory-modification by the introduction of a new allegedly content-increasing auxiliary hypothesis. This legitimation was to rule out auxiliary hypotheses that are *ad hoc* in the pejorative sense.

When discussing the 'immunization' of a core theory from a falsification by the legitimate introduction of a modifying auxiliary, Popper cites the paradigm example of the postulation of the planet Neptune, which served to explain observations of the motion of Uranus that had been incompatible with the prior version of Newtonian planetary theory.[22] As is apparent from his account there of this example and from his illustrative mention of other auxiliary hypotheses, Popper's stated requirement for the admissibility of an auxiliary is typically addressed to the following kind of logical situation: Existing evidence E is logically incompatible with a theory T_1, and a theory T_2 generated from T_1 by a modifying auxiliary hypothesis H is then able to explain E.[23]

Prima facie, one might be tempted to suppose quite mistakenly that these two versions of Newtonian planetary theory have comparable contents. Indeed, the avoidance of the serious error inherent in this supposition is a salutary antidote to Swinburne's sanguine assurance of content comparability, and to Popper's mishandling of this case.

Popper makes use, in addition to Tarski's notion of the logical content of a statement, of the following other sorts of content: (1) The 'informative content' $IC(T)$ of a theory T, which is the set of statements incompatible with T, and (2) the 'empirical content' $EC(T)$, which is the set of *observation* sentences that *contradict* T. And as he pointed out, for any two theories T_1 and T_2, the informative content $IC(T_1)$ of T_1 will be a subset of the informative content $IC(T_2)$ of T_2, if and only if the corresponding subset relation holds between their logical (Tarskian) contents $LC(T)$.[24] Formally expressed:

$$(T_1)(T_2) \, [\{IC(T_1) \subset IC(T_2)\} \leftrightarrow \{LC(T_1) \subset LC(T_2)\}].$$

[21] Op. cit. note 13, 6–9, 21–22.
[22] K. R. Popper, 'Intellectual Autobiography and Replies to My Critics', *The Philosophy of Karl Popper*, in Library of Living Philosophers, Book 1 (Chicago: Open Court, 1974), 32.
[23] K. R. Popper, *The Logic of Scientific Discovery* (London: Hutchinson, 1959), 83.
[24] Op. cit. note 22, 18.

Adolf Grünbaum

Now apply this result to the special case of an original theory T_1 and a *contrary rescuing* theory T_2 generated from T_1 by an auxiliary hypothesis H. Since T_1 and T_2 are incompatible, $LC(T_1) \not\subset LC(T_2)$. And it then follows from Popper's result that the *informative* content $IC(T_1)$ of T_1 can *never* be a proper subset of $IC(T_2)$.

Now let τ_1 be the *pre*-Neptunian version of Newtonian planetary theory of *circa* 1820, while τ_2 is the Leverrier and Adams amended version of solar planetary theory which asserts the existence of the additional trans-Uranian planet Neptune. Then the auxiliary hypothesis H is the assertion that this additional planet exists *and* exerts non-negligible gravitational influences on the motion of Uranus. Furthermore, if U_1 is the observation statement asserting τ_1's *incorrect* prediction of the orbit of Uranus, and U_2 is the *contrary* observation statement codifying the actually 'observed' orbit of Uranus, then τ_2 entails U_2, which is our paradigmatic datum E above. And since τ_1 and τ_2 are just a special case of our incompatible ancestor and rescuing theories T_1 and T_2, we have at once that $IC(\tau_1) \not\subset IC(\tau_2)$ and $EC(\tau_1) \not\subset EC(\tau_2)$, i.e., the *informative* content of τ_1 is *not* a subset of the informative content of τ_2, and similarly for the corresponding *empirical* contents.

It is very surprising that Popper *explicitly denied* both of these patent results when he wrote as follows:

> a prima facie falsification *may* be evaded ... as in the Uranus/Neptune sort of case, by the introduction of testable auxiliary hypotheses, so that the empirical content of the system—consisting of the original theory plus the auxiliary hypothesis—is greater than that of the original system. We may regard this as an increase of informative content ... There are, of course, also auxiliary hypotheses which are merely evasive immunizing moves. They decrease the content.[25]

It will be noted that here Popper asserts *incorrectly* that the empirical content of τ_1 is a subset of the empirical content of τ_2 and similarly for the corresponding informative contents, i.e., $EC(\tau_1) \subset EC(\tau_2)$, and $IC(\tau_1) \subset IC(\tau_2)$. To see how he was led to these false conclusions, observe that he uses the word '*plus*' when he characterizes the theoretical system τ_2 as 'consisting of the original theory $[\tau_1]$ plus the auxiliary hypothesis $[H]$'. His use of the word '*plus*' here seems to have misled him as follows: He logically misconstrued the so-called 'introduction' of the hypothesis H of an additional planet as a *mere conjunctive appending* of H to τ_1, thereby

[25] Op. cit. note 22, 33, item *e*.

270

inferring that τ_2 is the conjunction of τ_1 and H! In this way, he was apparently led to make the untenable assertion that the transition from τ_1 to τ_2 involved an *increase* of empirical and hence informative content. But this claim runs afoul of the following facts: (1) Far from generating τ_2 by being *appended conjunctively* to τ_1, the full-blown auxiliary H *replaces* the *contrary* constituent of τ_1 which (tacitly) asserts the *non*-existence of any trans-Uranian planet capable of exerting appreciable gravitational forces on Uranus, and (2) τ_1 and τ_2 are logically incompatible, if only because they respectively entail the contrary 'basic' orbital statements U_1 and U_2. Having apparently misdepicted τ_2 as the conjunction $\tau_1 \cdot H$, it is not surprising that Popper feels entitled to claim an increase of *IC* and *EC*.

Thus, content-incomparability aborted Popper's attempt to legitimate theory-modification by the introduction of a new content-increasing auxiliary hypothesis.

Case 4. As we saw, competing theories are not content-comparable. But conflicting theories are also commonplace in the history of science. How then is Swinburne going to implement the use of simplicity as a **tie-breaker** between theories whose contents he imperatively needs to be equal? We can match the demonstrated content-incomparability of Newton's and Einstein's theories by typical examples of pairs of incompatible theories that are less ambitious than they are and yet conceptually disparate and ontologically incongruous.

The following examples are in point: (i) The phlogiston versus Lavoisier's oxidation theory of chemical combustion. Each theory employs some essential concepts not present in the other; thus, even the notion of 'calx' just means ash in the phlogiston theory, while meaning 'oxide of a metal' for Lavoisier; (ii) The rival caloric and classical thermodynamic theories of heat in one of which caloric is a weightless actual fluid substance. They give diverging accounts of temperature-equalization between hot and cold; (iii) Newton's particle theory of light versus Huyghens's wave theory generate divergent questions; thus, Newton's theory could not readily deal with a question as to the wave length of monochromatic green light, (iv) Even in the case of Ptolemy's and Copernicus's theories, there are well-posed questions in the one theory, which are ill-posed in the other by respectively resting on presuppositions that are declared to be false in the other: e.g., Ptolemy can ask how long it takes for the sun to go around the earth, but Copernicus cannot; and Copernicus can ask how long it takes the earth to go around the sun, but Ptolemy cannot.

Adolf Grünbaum

How, one needs to ask, would Swinburne compare the contents within each pair of these theories as to equality, with a view to then comparing the overall simplicity of each of them? And, if the members of each pair are not certifiable as having equal content, what relevance do overall simplicity ratings have to the presumption that one of the two theories in the pair is more likely to be true than the other? After all, Thales's ancient *monistic* hydrochemistry is staggeringly simpler than Mendeleyev's 19th century *polychemistry*, but the latter is overwhelmingly more likely to be true. It emerges that Swinburne's recipe is epistemically hollow.

It would seem, therefore, that, despite his aforecited disclaimer concerning the assessment of the probability of rival scientific theories, it was altogether utopian on his part to tell us: 'I aim to show the nature and force of the criterion of simplicity by considering cases of conflicting theories where the other criteria for choice between them [such as parity of content] are equally well satisfied'.[26] Some of these other criteria are 'equal content [,] fitting equally well with background knowledge and yielding the data equally well'.[27] Alas, Swinburne engaged in special pleading by being content not to supply a single significant scientific illustration from *extant* 'cases of conflicting theories' differing in simplicity but satisfying equally well all of the other criteria for choice between them, such as equality of content.

The closest he comes to such an illustration is a design in which he *decrees*, merely programmatically without any specificity, the construction of theories of *equal content* as part of a presumed consensus among working scientists as follows:

> Take some standard scientific theory T_1; produce another theory T_2 which yields the data equally well *and has equal content*, and has some marginal advantage over T_1 in respect of one facet of simplicity at the expense of disadvantage in respect of another [italics added]. It will normally be evident to any scientist (even if he knows nothing about the particular field in question) which is the simpler theory and the one to be adopted if they fit equally well with background knowledge. But there certainly remains a borderland of pairs of formulations of theories about the overall simplicity of which scientists will differ.[28]

[26] Op. cit. note 3, 15.
[27] Op. cit. note 3, 23.
[28] Op. cit. note 2, 30–31.

It is very regrettable indeed that Swinburne's recipe here is very contrived, if only because his envisioned scenario is hardly a typical case of theory-competition. How, for example would his prescription here assure that the conflicting theories T_1 and T_2 have the same content? Moreover, he offers no historical data from any science, but only undocumented sociological *speculations*, about the comparative simplicity ratings made by scientists, when they are so inclined. Hence we don't know whether they will differ as to overall simplicity ratings more often than not, rather than only in a borderland of theories, as Swinburne would have it.

Summary

Karl Popper held that theories could be rank-ordered with respect to content, most notably by erotetic content in the case of incompatible theories such as Newton's and Einstein's theories of gravitation. But, as we saw, Popper's content-rankings were fundamentally unsuccessful and cast grave doubt on the content-comparability of rival theories., even when only modifying auxiliaries are introduced à la Neptune/Uranus

Swinburne claimed that simplicity is a *tie-breaker* among *conflicting* hypotheses as follows 'Simplicity is the only criterion of choice among hypotheses of *equal scope* [*content*] with equal ability to yield the data, when there is no background evidence ...' [italics added].[29] But, although the simplicity ratings avowedly pertain to rival hypotheses of *equal content*,[30] Swinburne has left the implementation of this requirement glaringly unfulfilled. And he did so after contenting himself with an *ipse dixit* devoid of even a single scientific example: 'rough comparisons [of content or "scope"] at any rate, and sometimes precise comparisons, are possible between theories' not related by one-way entailment.[31]

Therefore, I conclude that Swinburne has not given us a viable prescription for using the greater simplicity of one of two rival theories as evidence of its being more likely to be true.

[29] Op. cit. note 5, 93.
[30] Op. cit. note 3, 30.
[31] Op. cit. note 5, 82.

Adolf Grünbaum

Coda on Atheism versus Theism

In Swinburne's (2004, p. 336), he wrote: '... if there is to exist something, it seems impossible to conceive of anything simpler (and therefore *a priori* more probable) than the existence of God'. Furthermore, he told us that '... the [explanatory] choice is between the universe as [explanatory] stopping point [i.e., as existing qua brute fact] and God as [explanatory] stopping point [i.e., existing as a matter of brute fact]' (ibid., p. 147).

Thus the God hypothesis is supposedly the simpler option. And the avowed cardinal thesis of Swinburne's (2004) book is 'an argument for God being the cause [*ex nihilo*] of the existence of the universe' (ibid., p. 138, n. 9).

Just for argument's sake, posit with Swinburne that in the class of all existential hypotheses, the theistic one is the simplest, and therefore simpler than its atheistic competitor. It is then a corollary of the conclusion of our *Summary* above that this supposed greater simplicity does **not** show theism to be inductively more likely to be true as an explanation of the universe than an atheistic scientific account of it, absent the demonstration of the requisite *content-equality* of these two rival hypotheses.

Moreover, in my (Grünbaum, 2004), I have argued in very great detail contra Swinburne that (i) The very existence of the universe and of its laws does not require external causal explanation at all, because the demand for such explanation rests on multiply ill-founded assumptions, and (ii) the theistic explanations offered by Leibniz and Swinburne within the framework of these assumptions fail multiply to transform scientifically postulated brute facts into *explained* results.

References

Grünbaum, Adolf. 1973. *Philosophical Problems of Space and Time*, 2d ed. Dordrecht: D. Reidel Publishing.

—. 1976a. Can a Theory Answer more Questions than one of its Rivals? *British Journal for the Philosophy of Science* 27(1), pp. 1–23.

—. 1976b. Is the Method of Bold Conjectures and Attempted Refutations *Justifiably* the Method of Science? *British Journal for the Philosophy of Science* 27(2), pp. 105–136.

—. 1976c. Ad Hoc Auxiliary Hypotheses and Falsificationism. *British Journal for the Philosophy of Science* 27(4), pp. 329–362.

—. 2004. The Poverty of Theistic Cosmology *British Journal for the Philosophy of Science* 55(4), pp. 561–614.

Popper, K. R. 1959. *The Logic of Scientific Discovery*. London: Hutchinson.

—. 1972. *Objective Knowledge*. Oxford: Clarendon Press.

—. 1974. Intellectual Autobiography and Replies to My Critics. In *The Philosophy of Karl Popper*, ed. P. A. Schilpp, Library of Living Philosophers, Chicago: Open Court Publishing.

Swinburne, Richard. 1997. *Simplicity as Evidence of Truth*. Milwaukee, WI: Marquette University Press.

—. 2000. Reply to Grünbaum. *British Journal for the Philosophy of Science* 51(3): pp. 481–485.

—. 2001. *Epistemic Justification*. Oxford: Clarendon Press.

—. 2004. *The Existence of God*, 2d ed. Oxford: Clarendon Press.

ADOLF GRÜNBAUM
Center for Philosophy of Science
University of Pittsburgh
Pittsburgh, PA. 15260-2510, U.S.A.
E-mail: grunbaum@pitt.edu
Website: www.pitt.edu/~grunbaum/

Index

Index